The Yeast Connection and Women's Health is a valuable resource that provides a complete guide to understanding the relationship between yeast and human illness. So many women who have been told "you just have to live with it" will now find new direction and hope. This is a book every woman deserves to have in her library.

—James H. Brodsky, M.D.
Internal Medicine
Georgetown University Medical Center

Dr. Crook's insight and wisdom have changed the lives of thousands of my patients. I recommend his books to all who have been frustrated with their lack of health and energy and their recurrent illnesses. Now with Dr. Dean's update you will be informed and empowered to have the health you have been searching for.

—Jesse Lyne Hanley, M.D.
Women's Health Expert and Author of *Tired of Being Tired* and *What the Doctor May Not Tell You About Premenopause*

In addition to providing clear, practical advice to women puzzled by their lingering health problems, *The Yeast Connection and Women's Health* will hopefully serve as a valuable resource for doctors, nurses and other health professionals.

—Devra Davis
World-renowned Epidemiologist and Author of *When Smoke Ran Like Water* (Basic Books, 2002)

"A portion of the proceeds from this book will go to support organizations and initiatives that help women care for themselves, physically, emotionally, financially, and spiritually."

THE YEAST CONNECTION AND WOMEN'S HEALTH

Ms. Rosaria Chantrill
36 Woodlawn Ave
Auburn, ME 04210-4546

THE YEAST CONNECTION AND WOMEN'S HEALTH

WILLIAM G. CROOK, M.D.
WITH CAROLYN DEAN, M.D., N.D.
AND ELIZABETH B. CROOK

PROFESSIONAL BOOKS
Jackson, Tennessee

DISCLAIMER

This book describes relationships which have been observed between the common yeast germ *Candida albicans* and health problems which affect people of all ages and both sexes—especially premenopausal women. *We have written it to serve only as a general informational guide and reference source for both professionals and nonprofessionals.*

For obvious reasons we cannot assume the medical or legal responsibility of having the contents of this book considered as a prescription for anyone.

Treatment of health disorders, including those which appear to be yeast connected, must be supervised by a physician or other licensed health professional. Accordingly, you and the professional who examines and treats you must take the responsibility for the uses made of this book.

Published by Professional Books, Inc., Box 3246, Jackson, Tennessee 38303.

Cover art by Alternative Strategic Advertising and Promotions, Nashville, Tennessee. Book design and typography by ProtoType Graphics, Inc., Nashville, Tennessee.

Library of Congress Catalog Card Number: 95–067424

ISBN: 0-933478-27-5

Manufactured in the United States of America
2 3 4 5 6 7 — 09 08 07 06

Dedicated to the vast numbers of women whose courage and persistence have helped them regain their health

Table of Contents

Foreword

No one has done as much to educate the American people and the medical profession about yeast-related health problems as Dr. William Crook. Over the past twenty years he was a tireless advocate, traveling to meeting after meeting, interviewing clinicians, researchers and patients, and writing incessantly. Although Billy was a seasoned clinician before he became an author, during the past two decades he referred to himself as a medical journalist trying to give his readers as much useful information from as many different perspectives as possible. He always recognized the importance of creating a dialogue about yeast between his colleagues and readers and the mainstream medical profession. He was besieged by letters from readers who were unable to find doctors who would take the yeast connection seriously, and he knew that ignorance about yeast-related health problems among the medical profession was preventing huge numbers of people from receiving appropriate health care.

This important last work of Dr. Crook's provides much useful and practical information for women struggling with chronic health problems. Because you will most likely need the help of a health professional in implementing a comprehensive treatment program, I want to emphasize the importance of communicating your ideas and the information you learn here to your doctor in a fashion that makes sense to him or her. I have prepared a list of objections or statements that doctors frequently make about yeast. The statements in themselves are factually true, but they actually prevent the doctor from understanding the nature of the problem, so each statement is followed by a response. The objections and the responses tell a story.

1. "Everyone has yeast."

True. Almost all people are colonized by yeasts, which live in the mouth, the intestines and the vagina and can frequently be found in the nose and ears and on the skin. Yeasts only make people sick under special circumstances, which usually require an increase in the number of yeasts in some part of the body. Overgrown yeasts may make people sick by three mechanisms: *invasion, allergy* and *toxicity.* There has been a great deal of research concerning each of these mechanisms, but most doctors only think about *invasion,* which is the burrowing of yeast into the tissues of some part of your body. In the vagina, this is called "candida (or monilia) vaginitis," in the mouth or throat it is called "thrush," in the skin it is called "cutaneous candidiasis" or sometimes "tinea." If yeast actually invade the internal organs, like the liver, kidneys or heart, the condition is called "disseminated candidiasis."

2. "There is no way that you could have a systemic yeast infection. That is ridiculous."

True. When a doctor hears the term "systemic candida" or "systemic yeast infection," he or she thinks of disseminated candidiasis. This is an uncommon, often fatal disorder that occurs primarily in cancer patients on chemotherapy or in intravenous drug abusers. Even people with AIDS, who are prone to thrush and other forms of locally invasive candiasis because of their weakened immune systems, do not get disseminated candidiasis. If you feel sick all over or if you feel as if yeast has invaded your brain or messed up your hormones, it is not because of "systemic candidiasis" as your doctor understands it. It may be because of the other mechanisms by which yeast makes you sick: *allergy* or *toxicity.* These mechanisms do not even require that invasive yeast infection be present in your body. Therefore, knowing that you have an allergic or toxic reaction to yeast can be hard to demonstrate with medical tests.

3. "Everyone is allergic to yeast. That's normal."

True. Over ninety percent of a healthy population shows a type of allergic reaction to yeast called "delayed hypersensitivity" (DHS), which is also called "Type 4 allergy." In fact, the DHS response to yeast is what helps the body prevent yeast overgrowth. DHS responses to yeast are normal and protective. In my experience, many women with recurrent vaginal yeast infection show a lack of DHS to Candida upon skin testing and this lack of DHS is one reason why they have so much trouble preventing or clearing the yeast infection. There are other types of allergic reactions to yeast, however. These are not normal or common and can make you quite ill. One of these is called "Type 1 allergy" or "immediate hypersensitivity" and it is found in no more than ten per cent of the general population. Controlled clinical studies have shown that Type 1 allergy to Candida can cause chronic vaginitis, hives, asthma, eczema, diarrhea and abdominal cramps. Other types of unusual or abnormal allergic reactions to yeast may trigger autoimmune diseases like arthritis or thyroiditis or celiac disease (an intestinal disease due to an abnormal immune response to gluten, the major protein found in wheat). A study recently published in the leading British journal, the *Lancet,* showed that Candida contains a protein called HWP-1, which is similar in its structure to gluten. Candida infection in the intestines can cause an immune system reaction to HWP-1, which then stimulates an allergic reaction to the gluten in wheat and may trigger celiac disease in genetically susceptible people. If removing wheat from the diet doesn't control the symptoms of a person with celiac disease, it may be necessary to eliminate yeast from the intestinal tract with anti-fungal medication. (Nieuwenhuizen WF, Pieters RHH, et al. "Is *Candida albicans* a Trigger in the Onset of Coeliac Disease?" *Lancet,* June 21, 2003;361: 2152–2154.)

4. "I have never heard of toxicity caused by yeast infection."

True. Most doctors are unaware of the research because there are many different types of toxins made by yeast and they have many different kinds of effects. The most studied is a substance called zymosan, which is found in the capsule of the yeast. Zymosan causes inflammation. People suffering from psoriasis have abundant yeast growing in the affected skin. Research done by Dr. William Rosenberg and his colleagues at the University of Tennessee in Memphis has found that zymosan from these yeast cells causes much of the inflammation associated with psoriasis. Another toxic substance produced by yeast is arabinitol. People with yeast overgrowth may have high levels of this substance or its derivatives in blood and urine. Arabinitol is known to produce toxic effects on the brain and nervous system and the immune system of animals. There are numerous other toxins produced by yeasts that may account for fatigue, depression and hormone disturbances experienced by so many women with yeast problems. This is an area of budding research. Doctors who are not close to the research may know nothing about it. But the treatment of toxic yeast overgrowth is straightforward and explained in detail throughout all Dr. Crook's writings. If you have a complex illness that has not responded to conventional treatments and that may be yeast-related, aren't you entitled to a trial of therapy?

Leo Galland, M.D.
Director, *Center for Integrated Medicine*
New York City
July 2003

Introduction

If I told you that there is a fungus trapped in your intestines that produces 180 chemical toxins that are capable of shutting down your thyroid, throwing your hormones off balance and causing you to crave sugar and alcohol—all of which would also make you put on weight—would you believe me?

Chances are you would find all of this hard to believe. If this is true, why isn't this devastating condition and its cascade of consequences making headlines? At the same time, why isn't every high-tech research lab in the country frantically searching for a cure? Yet this problem, faced by millions of women, *is* kept under wraps, hidden in plain sight.

Many physicians are unaware of the extent of the problem or what to do about it. Many have even told women that their yeast-related problems are "all in your mind." In reality, the yeast problem is so extensive and so devastating that we have updated this book and included a completely new section on excess weight and the yeast connection.

All of this expands on the life's work of one extraordinary man. Dr. William Crook, "Billy" as his friends knew him, was an unusually kind and caring physician, exactly the type of physician you'd be delighted to find after shuffling from doctor to doctor in search of answers to your health problems. These may be problems that have plagued you, perhaps for years, with no one offering you the solutions you need—and which *are* available—as I said above, "hidden in plain sight."

Billy Crook was a humble and unassuming man. He withstood a great deal of professional ridicule and ostracism, which began in 1983 when he published *The Yeast Connection,* his first book on candida yeast and its potential for widespread destruction in the human body. Conventional medicine did not, and unfortunately still does not, fully understand the phenomenon of yeast overgrowth. However, doctors like Billy Crook persevered and brought hope to thousands of people desperate for answers.

In my years of conventional medical training, I was taught only that yeast causes vaginitis or it contributes to a life-threatening blood infection found in patients with advanced cancers or AIDS. Fortunately, during my naturopathic training in the early 1980s, I learned more about yeast overgrowth. Thus, from the time I entered practice, I treated yeast in my patients, often with remarkable results.

One day in 1986, I had the honor of being a guest on a Canadian television show, along with Dr. Crook. The 90-minute program was all about yeast. Astonishingly, the station tabulated 80,000 people trying to call in, all clamoring to ask questions. Talk about an unmet need!

After that show, I made sure that the treatment of yeast played an even greater role in my medical practice. By 1991, I had treated more than 2,000 cases and had begun to call yeast overgrowth "Crook's Candidiasis."

So what *is* this Crook's Candidiasis? In the simplest of terms, this condition involves an overgrowth of yeast that has been stimulated by the excessive intake of sugar and the overuse of antibiotics. These two factors cause yeast to invade the GI tract and create leaky gut syndrome, as well as a yeast allergy and other widespread adverse reactions. (These terms will be part of your vocabulary by the time you've finished this book.)

The insidious increase in yeast-related medical conditions is directly related to the advent of antibiotics some 60 years ago. I'm not saying that antibiotics are a bad thing. However, the pervasive use of antibiotics for viruses, for which they are completely ineffective, and in our food supply has caused a condition known as dysbiosis—an imbalance of good and bad flora in the intestines.

Billy Crook was not seeking to have a medical condition named after him, although I believe everyone should use the term "Crook's Candidiasis" as a tribute to his lifelong research, treatment and educational efforts on behalf of sufferers.

It was Dr. Crook's life's dream for this condition to be accepted by the medical community, and I'm honored to continue to carry his torch, to help this dream become a reality and to help millions of women with nowhere to turn.

Please enjoy this latest edition of *The Yeast Connection and Women's Health,* including the new eye-opening section on weight problems and yeast overgrowth.

This updated book is dedicated to the memory of the remarkable

Dr. Billy Crook, and to our readers who have the intelligence and perseverance to take control of their own well-being and find the solutions they need to bring them back to vibrant health.

Carolyn Dean, M.D., N.D.
September, 2005

Acknowledgments

I want to thank my father, William Crook for the manuscript of this book. This edition of *The Yeast Connection and Women's Health* is a result of the efforts of many talented and dedicated people.

Before my father's death in October 2002, he had written an acknowledgment thanking many people whom I would like to acknowledge, too.

He wrote, "To all my physician colleagues who provided me with new information for this book as well as other professionals and non-professionals who helped me in various ways, and to my effective and competent staff, Janet Gregory, Brenda Harris, and Jan Torre, I am deeply grateful."

In this and the previous edition, I have been fortunate to rely on the expertise of many talented people, especially Carolyn Dean, M.D., N.D. She is knowledgeable, compassionate, and persistent in reaching out to help the millions of people who suffer with yeast- related health problems. I appreciate her work for this edition. It would not be possible without her.

I'm grateful to Leo Galland, M.D., an exceptional physician who has reviewed the book and made thoughtful additions.

It has been a blessing to rely on the professional expertise of Meredith Carr, Judy Katz, Martica Griffin, and Kathleen Barnes for editing and design, John Adams and Jennifer Zimmerman for book design and typesetting.

Many others have provided invaluable assistance including Janet Gregory, my father's long time office manager and Joni Kane, my assistant. Finally I am indebted to my mother, Betsy Crook and my sisters Nancy Ward and Cynthia Crook for their support in innumerable ways. Thank you.

Elizabeth Crook
Nashville, Tennessee
July 2005

PART 1

Overview

CHAPTER 1

Are Your Problems Yeast Connected?

I f you . . . feel sick all over,

- have taken antibiotic drugs,
- have sought help from many different specialists,
- have taken birth control pills,
- are troubled by fatigue and/or muscle aches,
- are bothered by food sensitivities and/or digestive problems,
- crave sugar,
- sometimes feel spaced out,
- are bothered by depression and/or irritability
- are bothered by headaches
- are sensitive to tobacco smoke, perfume and other chemicals . . .

your health problems may be yeast-connected!

Women between the ages of 20 and 55 are especially apt to develop yeast-related health problems. Common symptoms include:

- weight gain
- recurrent vaginal yeast infections
- recurrent urinary tract infections
- PMS
- vulvodynia (burning vulva)
- sexual dysfunction (loss of interest in sex or loss of orgasm)
- dyspareunia (painful intercourse)

3

- infertility
- endometriosis
- interstitial cystitis

Men and children also develop yeast-related health problems, and these are especially common in people who have taken many antibiotic drugs.

Other complaints and illnesses that may sometimes be related to the common yeast, *Candida albicans,* include:

- numbness
- tingling
- multiple sclerosis
- Crohn's disease
- irritable bowel syndrome
- colitis
- rheumatoid arthritis
- myasthenia gravis
- eczema
- acne
- lupus erythematosus
- asthma
- sinusitis
- psoriasis
- chronic hives

 See www.yeastconnection.com for a printable monthly chart to help you identify and track your symptoms.

I'm not saying that the common yeast, *Candida albicans* or any of its close relatives, are the cause of all these problems. Yet candida may be one of the causes—even a major cause—of these and other health-related problems.

In this book, you'll find easy-to-read-and-understand explanations of yeast-related health problems and a step-by-step program that will help you overcome them.

CHAPTER 2

Candida Questionnaire and Score Sheet

I have been impressed by the remarkable uniformity of the histories I have found in my patients with yeast-related problems.

While physical and laboratory examinations provide a physician with some of the clues that lead to a diagnosis, I've found that this questionnaire is perhaps my most valuable tool in finding a way to help patients who feel "sick all over."

This questionnaire is designed for adults and is not appropriate for children. Since your spouse, companion, friend or male relative may also be troubled by yeast-related problems, this questionnaire is designed for people of both sexes.

Section A focuses especially on factors in your medical history that promote the growth of *Candida albicans.*

Section B has a list of 23 symptoms usually present in people with these problems.

Section C consists of 33 other symptoms that are sometimes seen in people with yeast problems—and that are found in people with other disorders. Your total score on this questionnaire should help you and your physician evaluate the role *Candida albicans* plays in your problems. While it will not provide an automatic "yes" or "no" answer, it will provide some important clues.

SECTION A: HISTORY

	Point Score
1. Have you taken tetracycline or other antibiotics for acne for one month (or longer)?	35
2. Have you at any time in your life taken broad-spectrum antibiotics or other antibacterial medication for respiratory, urinary or other infections for two months or longer, or in shorter courses four or more times in a one-year period?	35
3. Have you taken a broad-spectrum antibiotic drug—even in a single dose?	6
4. Have you, at any time in your life, been bothered by persistent prostatitis, vaginitis or other problems affecting your reproductive organs?	25
5. Are you bothered by memory or concentration problems—do you sometimes feel spaced out?	20
6. Do you feel "sick all over" yet, in spite of visits to many different physicians, the causes haven't been found?	20
7. Have you been pregnant . . . two or more times? one time?	5 3
8. Have you taken birth control pills . . . For more than two years? For six months to two years?	 15 8
9. Have you taken steroids orally, by injection or inhalation . . . for more than two weeks? For two weeks or less?	15 6
10. Does exposure to perfumes, insecticides, fabric shop odors and other chemicals provoke . . . Moderate to severe symptoms? Mild symptoms?	 20 5

11. Does tobacco smoke really bother you?	10
12. Are your symptoms worse on damp, muggy days or in moldy places?	20
13. Have you had athlete's foot, ringworm, "jock itch" or other chronic fungus infections of the skin or nails? Have such infections been . . .	
Severe or persistent?	20
Mild to moderate?	10
14. Do you crave sugar?	10

Total Score, Section A _____

SECTION B: MAJOR SYMPTOMS

For each of your symptoms, enter the appropriate figure in the Point Score column:

> If a symptom is occasional or mild, 3 points
> If a symptom is frequent and/or moderately severe, 6 points
> If a symptom is severe and/or disabling, 9 points

Add total score and record it at the end of this section.

Point Score

1. Fatigue or lethargy _____

2. Feeling of being "drained" _____

3. Depression or manic depression _____

4. Numbness, burning or tingling _____

5. Headaches _____

6. Muscle aches _____

7. Muscle weakness or paralysis _____

8. Pain and/or swelling in joints   _____

9. Abdominal pain   _____

10. Constipation and/or diarrhea   _____

11. Bloating, belching or intestinal gas   _____

12. Troublesome vaginal burning, itching or discharge   _____

13. Prostatitis   _____

14. Impotence   _____

15. Loss of sexual desire or feeling   _____

16. Endometriosis or infertility   _____

17. Cramps and/or other menstrual irregularities   _____

18. Premenstrual tension   _____

19. Attacks of anxiety or crying   _____

20. Cold hands or feet, low body temperature   _____

21. Hypothyroidism   _____

22. Shaking or irritability when hungry   _____

23. Cystitis or interstitial cystitis

Total Score, Section B   _____

SECTION C: OTHER SYMPTOMS

For each of your symptoms, enter the appropriate figure in the Point Score column:

If a symptom is occasional or mild, 1 point
If a symptom is frequent and/or moderately severe, 2 points
If a symptom is severe and/or disabling, 3 points

Add total score and record it at the end of this section.

Point Score

1. Drowsiness, including inappropriate drowsiness _____

2. Irritability _____

3. Incoordination _____

4. Frequent mood swings _____

5. Insomnia _____

6. Dizziness/loss of balance _____

7. Pressure above ears . . . feeling of head swelling _____

8. Sinus problems . . . tenderness of cheekbones
   or forehead _____

9. Tendency to bruise easily _____

10. Eczema, itching eyes _____

11. Psoriasis _____

12. Chronic hives (urticaria) _____

13. Indigestion or heartburn _____

14. Sensitivity to milk, wheat, corn or other common
    foods _____

15. Mucus in stools _____

16. Rectal itching _____

17. Dry mouth or throat _____

18. Mouth rashes, including "white" tongue _____

19. Bad breath _____

20. Foot, hair or body odor not relieved by washing _____

21. Nasal congestion or postnasal drip _____

22. Nasal itching _____

23. Sore throat   _____

24. Laryngitis, loss of voice   _____

25. Cough or recurrent bronchitis   _____

26. Pain or tightness in chest   _____

27. Wheezing or shortness of breath   _____

28. Urinary frequency or urgency   _____

29. Burning during urination   _____

30. Spots in front of eyes or erratic vision   _____

31. Burning or tearing eyes   _____

32. Recurrent infections or fluid in ears   _____

33. Ear pain or deafness   _____

Total Score, Section A   _____

Total Score, Section B   _____

Total Score, Section C   _____

GRAND TOTAL SCORE   _____

The Grand Total Score will help you and your physician decide if your health problems are yeast connected. Scores in women will run higher, as seven items in the questionnaire apply exclusively to women, while only two apply exclusively to men.

Yeast-connected health problems are almost certainly present in women with scores over 180, and in men with scores over 140.

Yeast-connected health problems are probably present in women with scores over 120, and in men with scores over 90.

Yeast-connected health problems are possibly present in women with scores over 60, and in men with scores over 40.

With scores of less than 60 in women and 40 in men, yeasts are less apt to cause health problems.

CHAPTER 3

Yeasts—What They Are and How They Make You Sick

WHAT ARE YEASTS?

Yeasts are single-cell living organisms which are neither animal nor vegetable. They live on the surfaces of all living things, including fruits, vegetables, grains and your skin. They're a part of the "microflora" which contribute in various ways to the health of their host.

Yeast itself is nutritious, and small amounts of yeasts give bread its light texture. Yeast is a kind of fungus. Mildew, mold, mushrooms, monilia and candida are all names that are used to describe different types of yeast. And one family of yeasts, *Candida albicans,* normally lives on the warm, inner creases and crevices of the digestive tract, vagina and skin.

Healthy women have a natural community of *Candida albicans* organisms that live in all three locations. When your immune system is healthy, friendly intestinal bacteria like *Bifidobacteria bifidum* and *Lactobaccillus acidophilus* create a symbiotic system with the *Candida albicans* yeast cells that keeps everything in balance. But when this system gets out of balance, *Candida albicans* yeast cells rapidly overwhelm the friendly bacteria and can result in an overload with potentially serious effects. This condition and conditions resulting from imbalances of intestinal flora is sometimes referred to as dysbiosis.

FACTORS THAT PREDISPOSE YOU TO YEAST INFECTIONS

Prescription Medications

Yeast infections are especially apt to trouble you if you have taken repeated or prolonged courses of amoxicillin, ampicillin, Ceclor, Keflex, tetracycline and/or other broad-spectrum antibiotics during infancy, childhood, adolescence or since you've become an adult.

Antibiotics don't kill candida yeasts. Accordingly, if you took these drugs for acne (or for urinary, sinus, ear, chest or other infections), candida multiplied in your intestinal tract—and also in your vagina because the natural balance had been disturbed.

Other medications prescribed by your doctor may play an important part in causing yeast infections—especially the cortisone group of drugs when taken by mouth or injection, or when sprayed into your respiratory tract.

Sugar and Other Simple Carbohydrates

Sugars of many types promote the multiplication of *Candida albicans* in the digestive tract and play a major role in causing all of the health problems affecting women that I've discussed in this book. Sugar quite literally feeds the yeast. Two of my consultants, gynecologists John Curlin and Don Lewis, have told me on many occasions about the effects of sweets on patients with dysbiosis: "If my patients control their intake of sweets and other simple carbohydrates, they'll do well. But, if they do not, their symptoms will return," says Dr. Lewis.

 See www.yeastconnection.com for a comprehensive list of tools and techniques for managing your yeast connection.

OTHER FACTORS

Nylon underwear and tights may make you more apt to develop genital yeast infections. Such infections are especially apt to occur if your skin or mucous membranes are irritated or broken. They may also occur in the mouths of people with dentures.

Hormonal changes, especially during the premenstrual phase of your cycle, also encourage yeast overgrowth and in the opinion of candida pioneer Dr. C. Orian Truss, women who take birth control pills are also more apt to be bothered by yeast-related problems. That has been my experience as well, although I am unsure why there seems to be a connection. However, two of my gynecologist consultants told me in 2002 that women may take some of the presently available birth control pills and not develop yeast problems—especially if they're controlling their diets and doing the other things they need to do to enjoy good health.

A YEAST INFECTION IS MUCH MORE THAN A VAGINAL INFECTION

In their scientific studies carried out over 25 years ago, Dr. Mary Miles and her colleagues at Michigan State University found that every woman with a vaginal infection had an accompanying overgrowth of yeast in her digestive tract.[1] Based on the observations of scores of women I saw in my practice in the early and mid-1980s and the thousands of women who have sent me letters and e-mails since then, many women develop yeast-related disorders even though they had not been bothered by vaginal problems.

What Is "The Yeast Connection"?

It is a term to indicate the relationship of superficial yeast infections in your vagina and digestive tract to fatigue, headaches, depression, PMS, irritability and other symptoms that can make you feel "sick all over."

Superficial yeast infections cause symptoms in distant parts of the body. Scientists say there are several possible reasons how this happens:

Japanese researchers suggest *Candida albicans* puts out a variety of types of toxins that can weaken your immune system.[2,3]

One Cornell researcher theorizes that the candida infection can actually weaken your system enough to give you a higher risk of reinfection and of hormonal dysfunction. The development of autoantibodies like these can lead to rheumatoid arthritis. But the same study showed antifungal medications can eliminate these conditions.[4]

In the 1980s, researchers in Finland carried out studies on the relationship of *Candida albicans* to atopic dermatitis (eczema).[5]

Absorption of Food Antigens and Toxins

Based on clinical and research studies by many different observers, candida overgrowth in the intestinal tract may create what has been called a "leaky gut." With leaky gut, the mucosal lining between the intestine and the bloodstream breaks down, allowing harmful toxins and allergens to enter the bloodstream. Once they enter the bloodstream, these toxins can rapidly travel to all parts of your body, where they can further weaken your immune system and produce a long list of symptoms.

As a result, food antigens and toxins may be absorbed, which play a part in making you feel "sick all over."

> I'll simplify this explanation because it's so important: Every person with a yeast-related problem has an overgrowth of *Candida albicans* in the digestive tract. This creates a disturbance in the normal balance of good bacteria, which, in turn, leads to a weakness of the membrane lining the intestinal tract. This is commonly called a "leaky gut." As a result, antigens (or toxic substances) in food are absorbed, and this plays a part in making you sick.

Allergies to the Candida Yeast Itself

A number of physicians say actual allergies to the candida yeast may play an important role in causing symptoms.[6]

Internal medicine specialist James H. Brodsky, M.D., internal medicine clinical instructor at Georgetown University Medical Center in Chevy Chase, Maryland, thinks *Candida albicans* is one of the most allergy-inducing substances known—and that short-term and long-term super sensitivity to candida is very common among adults.

"The investigators estimate that in about 26% of patients with chronic urticaria (hives), *Candida albicans* sensitivity is an important factor. Significant clinical improvement was seen with anti-candida therapy and a low-yeast diet,"[7] says Dr. Brodsky.

REFERENCES

1. Miles, M. R., Olsen, L. and Rogers, A., "Recurrent Vaginal Candidiasis: Importance of an Intestinal Reservoir," *JAMA,* 238:1836–1837, October 28, 1977.

2. Iwata, K. and Yamamoto, Y., "Glycoprotein Toxins Produced by *Candida albicans,"* Proceedings of the 4th International Conference on the Mycoses, June 1977, PAHO Scientific Publication No. 356.

3. Iwata, K. and Uchida, K., "Cellular Immunity in Experimental Fungus Infections in Mice: The influence of infections in treatment with a candida toxin on spleen lymphoid cells," Mykosen, Suppl. 1, 72–81 (1978), Symposium Medical Mycology, Flims, January 1977.

4. Witkin, S. S., "Defective Immune Responses in Patients with Recurrent Candidiasis," *Infections in Medicine,* May/June 1985, pp. 129–131.

5. Savolainen, J., Lammintausta, K., Kalimo, K. and Viander, M., *"Candida albicans* and atopic dermatitis," *Clinical and Experimental Allergy,* Vol. 23, 1993, pp. 332–339.

6. Liebeskind, A., *Annals of Allergy,* 1962; 20:394–396; Hosen, H., "Focal fungal infections treated by immunological therapy with emphasis on vaginal moniliasis," *Texas Medicine,* 1971; 67:58; Kudelko, N. M., "Allergy in chronic monilial vaginitis," *Annals of Allergy,* 1971; 29:266; Palacios, H. J., "Hypersensitivity as a cause of dermatologic and vaginal moniliasis resistant to topical therapy," *Annals of Allergy,* 1976; 37:110–113; and Truss, C. O., *J. of Ortho. Psych.,* 1980; 9:287–301.

7. Brodsky, J. H., as quoted in the foreword of *The Yeast Connection,* Third Edition, paperback, by W. G. Crook, M.D., Professional Books, Jackson, TN and Vintage Books, New York, 1986.

CHAPTER 4

Why You May Develop Yeast-Related Health Problems

If you're bothered by vaginitis, vulvodynia, PMS or endometriosis, you've developed these problems because your immune system isn't as strong as it should be.

Other disorders that may bother you, including weight gain, fibromyalgia, chronic fatigue, sexual dysfunction, sinusitis and asthma, also develop because your immune system has been weakened and a *Candida albicans* overgrowth may have occurred, resulting in what experts call *dysbiosis*. This is a state of imbalance in the microorganisms of the digestive tract that results in disease. Dr. Cass explains to her patients that dysbiosis means "The bad bugs in the gut have outnumbered the good bugs. The goal is to restore balance, i.e., get rid of the bad and increase numbers of good guys."

Chemicals, molds and nutritionally deficient diets are other factors that play an important part in causing your problems. The most important of these factors are antibiotics. This includes not only antibiotics you've taken during your adult life, but also antibiotics given to you when you were a young child or a teenager.

Many different factors play a part in making you sick. I'm convinced that repeated courses of broad-spectrum antibiotics are the main "villains." These antibiotics cause yeast overgrowth in your intestinal tract and vaginal yeast infections. These infections set off disturbances that can make you feel "sick all over."

These antibiotics, often given unnecessarily, wipe out friendly bacteria in your intestinal tract, vagina and respiratory membranes. As a result, the usually benign yeast *Candida albicans* multiplies.

Candida yeasts aren't killed by antibiotics . . . so they multiply and raise large families, each of which in turn raises its own large family.

Candida overgrowth causes:

- A weakening of your immune system;
- Disturbances of the membrane lining of your intestinal tract, creating what is called a "leaky gut."

As a result, food allergens and toxins are absorbed into your circulatory system and reach every part of your body.

When your immune system is weak, you may develop repeated minor respiratory problems, including coughs, colds and sore throats. Although these minor problems are caused by viruses, you may be given an antibiotic that you and your physician hope will help you get rid of your symptoms sooner.

Most physicians today discourage the overuse of antibiotics, although they are necessary—even essential—in treating streptococcal sore throat, pneumococcal pneumonia and other serious and potentially life-threatening bacterial infections.

Regardless of the reason, when you take antibiotics a vicious cycle may develop.

Repeated courses of antibiotics set up a cascade that leads to many other problems.

Still, other factors play a role in making you sick or causing you to develop other types of problems, including:

- Sugar and other simple carbohydrates:
    A diet rich in sugar and other simple carbohydrates literally feeds yeast and promotes overgrowth.

 See www.yeastconnection.com for a chart of foods and their sugar content.

In the late 1800s, the average American ate four pounds of sugar a year. A century later, our consumption of sugar was 160 pounds per person per year. And for the year 2000, projected intake of sugar was 200 pounds per person per year. This is a 50-fold increase in the amount of refined carbohydrates within 130 years. We're in serious trouble!

Other "culprits" behind yeast overgrowth include:

- Hormonal changes associated with normal the menstrual cycle
- Birth control pills
- Pregnancy
- Steroids by pill, injection or inhalation
- Genital irritations and abrasions
- Reinfection from your sexual partner
- Diabetes

Many of your yeast-related symptoms, including PMS, sexual dysfunction, headaches and depression develop because your immune system, your endocrine system and your brain are intimately connected, and although we sometimes forget it, every part of your body is connected to every other part.

On many occasions during the past 20 years, my friend and mentor Sidney M. Baker, M.D., a functional nutrition specialist in Weston, Connecticut, has said:

Labeling diseases isn't the way we should go. When a person is tired and suffers from other chronic complaints, it's important for the physician to ask these two questions:

- Is there something that this person needs that she is lacking?
- Is there something that she is getting too much of that contributes to her problem?

If your doctor doesn't ask you these questions, take the bull by the horns and raise these issues with your doctor yourself.

See www.yeastconnection.com for guidance on discovering what your body is telling you.

Dr. Baker continued:

Together, these two questions form the basis for detective work aimed at uncovering imbalances in people of all ages, with various problems. Things that you may lack include:

- a nutritious diet that features vegetables and a variety of other good foods
- nutritional supplements, including antioxidants, magnesium, B vitamins, zinc and essential fatty acids
- full-spectrum light, clean air and pure water
- love, praise, touch and other psychological nutrients
- exercise

The things you should avoid as much as possible are:

- pollutants in the air, food, soil and water (such as pesticides, tobacco smoke and odorous chemicals)
- nutritionally poor foods and beverages (such as packaged and highly processed foods loaded with sugar and partially hydrogenated fats)
- inhalant and food allergens
- harmful microorganisms, including yeasts, molds, bacteria, viruses and parasites.

CHAPTER 5

Why Women Are More Affected Than Men

Women develop yeast-related problems more often than men, and perimenopausal (before menopause) women appear to be especially susceptible. There are probably several reasons for this.

In his classic book, *The Missing Diagnosis,* Dr. C. Orian Truss clearly describes the unique problems of women with candida-related health problems. He pointed out that countless women between puberty and menopause are troubled by vaginal problems, PMS, digestive symptoms, personality changes, concentration problems and a destructive loss of self-confidence.

DIFFERENCES IN ANATOMY

Women are more likely than men to develop urinary and genital yeast infections because of their anatomy. For example, a woman's urethra (the tube leading from the urinary bladder to the outside) is short, making it easier for bacteria to enter the woman's bladder and set up an infection.

Urinary tract infections are especially apt to occur in women after frequent or prolonged sexual intercourse. These infections are usually treated with antibiotics, allowing the candida yeast that lives normally in the intestinal tract to multiply because the natural balance with friendly and unfriendly bacteria has been disrupted.

The short distance between the anal opening and the vulva and vagina increases a woman's chances of developing a genital infection.

Since yeasts thrive on the warm, dark interior membranes of the body, the vagina furnishes a hospitable home.

THE PILL

Experts say at least 35% of women using birth control pills are susceptible to repeated bouts of vaginitis caused by *Candida albicans* yeast and, by at least some estimates, the figure is as high as 50%.

HORMONAL CHANGES

Yeast colonization is encouraged by hormonal changes in the normal menstrual cycle, during pregnancy and when hormones begin to fluctuate with age. Some women begin to experience hormonal swings in their early 30s, but the majority of women experience these swings beginning in their early 40s.

The culprit may be progesterone, the hormone that is produced in very small amounts just before ovulation. During pregnancy, high levels of progesterone persist, and during perimenopause, hormonal levels may fluctuate several times a day. For reasons medical science does not yet understand, these higher levels of progesterone correspond to rapid yeast growth in women.

Cornell's Dr. Steven S. Witkin validated this theory in his 1991 research that showed progesterone actually stimulates the growth of *Candida albicans*.[1]

PERIMENOPAUSAL WOMEN GO TO DOCTORS MORE OFTEN THAN MEN

Women are more likely than men to develop personal relationships with a physician because women go more often for routine checkups and Pap smears, when they are pregnant or when they develop vaginal yeast infections.

Accordingly, when a woman develops a fever, cough or cold, she's more likely to contact her physician to ask for relief. The physician may then prescribe an antibiotic, promoting the growth of yeasts.

PROLONGED ANTIBIOTIC USE FOR TEENAGERS WITH ACNE

Teenagers, especially girls, are concerned about their complexions. That makes them more likely to consult a doctor and to be put on long-term antibiotics. Although these drugs can be helpful to some teenagers with acne, they wipe out normal bacteria in the intestinal tract. As a result, yeasts multiply and a cascade of other health problems can develop in the short- and long-term.

I have received tens of thousands of phone calls and letters, most of them from women between the ages of 25 and 50 with complaints of being "sick all over." In response to these calls and letters, here's one of the first questions I ask:

"Did you, during your teen years, take long-term tetracycline for acne?"

Many of these people, including men, answer, "Yes."

In my mind, this is a red flag indicating the problems could be yeast-related, so I advise them to look at the anti-yeast program that has been helpful to so many patients.

REFERENCE

1. Kalo-Klein, A., Witkin, S. S., "Regulation of the immune response to *Candida albicans* by monocytes and progesterone," American Journal of Obstetrics and Gynecology, May 1995, 164 (5 Pt 1) 1351–4.

CHAPTER 6

The Diagnosis of a Yeast-Related Disorder

How does your physician make a diagnosis? If you're bothered by any of the symptoms, particularly the women's disorders listed on pages 3 and 4, it's essential for you to have a diagnostic workup that includes:

- *Your complete medical history.* Such a history includes not only your present complaints, but also a very detailed medical history, beginning with infancy.

 See **www.yeastconnection.com** for a printable chart to help you record your medical history.

- *A physical examination.* It should include a gynecological examination and examinations of your skin, eyes, heart, lungs and other parts of your body. Your health professional should also really *look* at you. I always look carefully at my patients. Even people who say they are in reasonably good health often look pale with dark circles or bags under their eyes and/or a wrinkle across their nose. If they are not anemic, I am often able to connect these features to hidden food allergies that are usually caused by the person's favorite foods.
- *Laboratory examination.* Tests, including blood tests, urine tests, stool examinations, X-rays and more complex laboratory studies.

This will likely include a gastrointestinal evaluation panel for *Candida albicans*. With some health disorders, a diagnosis can be made easily. Here are examples:

You develop a slight sore throat, a cough and other mild respiratory symptoms. Then you sit through an entire football game on a cold, windy day. That night your cough worsens. Early the next morning your teeth begin to chatter and you shiver and shake. You're having a chill. Then your fever jumps up to 104 degrees and you experience a sharp pain in your chest. You go to a hospital emergency room. The physician there listens to your story and examines you. Then she orders a chest x-ray and lab tests of various sorts, including a white blood count and a urinalysis.

After reviewing these findings, she says, "You have lobar pneumonia. With a prescription of penicillin today, you'll be a lot better by tomorrow. Then, with a few days of rest and penicillin, you'll be well in a week."

Here's another example:

You develop urinary frequency and burning and note blood in your urine. A microscopic examination shows that your urine is loaded with blood cells, pus cells and bacteria. These findings show that you have an acute urinary infection that requires an antibiotic given according to the physician's directions.

Diagnosing yeast-related fatigue, depression and many other disorders is an entirely different situation. Here's why: A physical examination and tests usually are helpful in directing your physician along the road toward a diagnosis, but they frequently do not provide a clear-cut answer. They do not enable your physician to "make a diagnosis." This is why it is often a prolonged process to come to a diagnosis of candidiasis. Nevertheless, it's important for any woman who feels "sick all over" to go to a physician (and I hope that doctor will be kind and caring) for a careful examination to make sure her symptoms are not caused by another health disorder.

If your physical examination and routine laboratory tests show no significant abnormalities and your history suggests a yeast-related problem, the best way of making a diagnosis is by noting your response to a simple, but comprehensive, treatment program. Such a program features a sugar- and yeast-free diet, probiotics and prescription or non-prescription anti-yeast medications.

In my own patients who were troubled by severe or long-lasting health problems, I always put in the "first team"—a systemic antifungal prescription medication—usually Diflucan.

See www.yeastconnection.com for a printable packet of information and charts to help you prepare for your doctor's appointment.

In order to help your physician diagnose your illness, it's essential to provide a very detailed medical history. Here's some of the information you need to provide:

1. When and how did your symptoms begin?
2. Did your symptoms begin suddenly like being hit by a stray bullet?
3. Or did they come gradually over a period of weeks, months or years?
4. Have you had illnesses over your lifetime that have required repeated courses of antibiotics?
5. What do you eat for breakfast, lunch, dinner and snacks?

See www.yeastconnection.com for a printable weekly chart to help you monitor your food intake.

6. Do you smoke?
7. Do you drink alcoholic beverages?
8. What drugs or medications do you take?
9. What are your present environmental exposures at home or at work?
10. Are you bothered by:
    - perfumes and colognes?
    - fabric shop odors?
    - dusts or molds?

- pollens?
- animal danders?

11. Are you exposed to smoke, insecticides, a gas cooking stove or new carpets in your home or workplace?
12. Do you use lawn chemicals or pesticides outdoors?

Come to your appointment prepared to provide this information. However, this in no way is a substitute for a careful, one-on-one, private doctor-patient discussion. In my experience, more than anything else, people with yeast-related health problems want to be listened to by a kind, compassionate, empathetic physician.

Now for a few more words about tests: There are a number of helpful tests that can assist with a diagnosis of candidiasis, including:

1. Comprehensive Digestive Stool Analysis (CDSA) by the Great Smokies Diagnostic Laboratory in Asheville, NC. website: www. gsdl.com. (800) 522-4762
2. Candida blood studies of various types, including measurements of candida antibodies, antigens and immune complexes
3. Stool studies for candida and other organisms

In addition, the vast majority of people with yeast-related health problems suffer from food sensitivities. Although commonly used allergy scratch tests often show negative results, other tests may be helpful, including:

1. The ALCAT test: a blood test that measures sensitivities to food extracts, food additives, chemicals and molds
2. IgG food sensitivity tests for delayed-onset food allergies
3. Tests to assess intestinal permeability: urine collected over a six-hour period will help determine if you have leaky gut.

 See www.yeastconnection.com for a printable sheet of the tests and labs listed on these pages to include in your packet of information for your physician.

Labs that offer tests that can help your physician with evaluating your complex health problems are:

- AAL Reference Laboratory, Santa Ana, CA (800) 522-2611 website: www.antibodyassay.com.
- Accu-Chem Laboratories, Richardson, TX (800) 451-0116 website: www.accuchem.com.
- American Medical Testing Laboratories, Hollywood, FL (800) 881-2685, website: www.aml.com.
- Cerodex Lab, Washington, OK (405) 288-2383, website: www. immy.com.
- Consulting Clinical and Microbiology Laboratories, Portland OR (503) 222-5279
- Diagnos-Techs, Inc., Kent, WA (425) 251-0596, website: www. diagnostechs.com.
- Doctors Data Inc. Chicago, IL (800) 323-2784, website: www.doctorsdata.com.
- Immuno Laboratories, Ft. Lauderdale, FL (800) 231-9197, website: www.immunolabs.com.
- Immunodiagnostic Laboratory, San Leandro, CA (800) 888-1113
- Immunosciences Lab, Beverly Hills, CA (310) 657-1077, website: www.immuno-sci-lab.com.
- Meridian Valley Clinical Laboratory, Kent, WA (800) 234-6825, website: www.meridianvalleylab.com.
- Metametrix Research Laboratory, Norcross, GA (770) 446-5483, website: metametrix.com.

In the past couple of years, during 2001 and 2002, I've become increasingly impressed with the value of electrodermal (BioMeridian) testing, and I discuss this method of testing in Chapter 34 of this book.

CHAPTER 7

Psychological Factors

Psychological factors are important in people with every health problem, whether it is heart disease, arthritis, fatigue, headaches, depression or other symptoms commonly found in people with yeast-related problems. It is well documented that these factors weaken the immune system. For example, studies I've read in medical journals show that T-lymphocytes (the cells that protect people from viral and other infections) were significantly reduced following the death of a spouse or a child, making the person more susceptible to infection.

On a more positive note, the observations of the late Norman Cousins show that the immune system can be strengthened and recovery from many illnesses accelerated by laughter and other psychological nutrients.

During the 1980s, I visited Mr. Cousins in his office at UCLA. During our fascinating conversations, he told me of his work with groups of patients with arthritis, cancer and other chronic, disabling and painful illnesses. He said, "I always tell them jokes. After the third joke I've seen them laughing so hard they have to hold their sides. Then when my session is complete I ask, 'How many of you are still hurting as much as you were when you walked into this room?' No hands are raised."

Surgeon, writer and lecturer Bernie Siegel, who published a number of books about the importance of psychological support, stresses the important link between mind and body. He urges people with chronic diseases to learn to have fun. Specific suggestions included reading humorous books and going to light, entertaining movies. He also urges people to play games and tell jokes to their friends. He suggests you can have fun with coloring books and do anything else that will bring out the child inside you.

In his book, *Maximum Immunity,*[1] published more than a decade ago, Michael A. Weiner, Ph.D., described a study carried out on cadets at West Point to see how psychological factors influenced their suscepti- bility to infectious mononucleosis.

"Cadets were selected at the beginning of the study who were free of the antibody for the Epstein-Barr virus . . . During their stay at the academy, some of the young men developed the Epstein-Barr virus an- tibody; but only some of these actually developed mononucleosis. The others remained symptom-free, indicating that they had better resis- tance. The cadets who became sick with infectious mononucleosis were generally found to have experienced greater academic pressure and to have shown poorer academic performance than the resistant group of cadets."

Weiner cited another study of a group of students in which "it was found that failure, social isolation and unresolved role crisis was often associated with respiratory infections. The more serious the sickness, the more likely it was that stressful situations had occurred during the preceding year."

In his 1998 book, *Power Healing,*[2] Dr. Leo Galland discussed "the four pillars of healing." These pillars include diet, exercise, the environ- ment and getting rid of internal toxins that play a part in making people sick. His fourth pillar focuses on interpersonal relationships and how they can help people get well.

I especially liked his discussion of the qualities of a caring doctor that he said all competent physicians must possess. These include the ability to listen, willingness to acknowledge patients' ideas and feelings about their illnesses, ability to show empathy and willingness to offer encouragement, hope and assurance.

When people write or call me seeking a physician, here's one of the first questions I ask: "Is your personal health professional kind and car- ing?" If the answer is "Yes," I say, "She is the best person to help you, even if she knows little about yeast-related disorders."

You'll find a further discussion of Dr. Galland's approach on his website www.mdheal.org.

More than a decade ago, the late Dr. Norman Vincent Peale, a won- derful minister and dynamic speaker, wrote a fascinating book, *Imaging.* In the book, he emphasizes that reaching a goal, such as regaining your health, takes more than simply thinking about it. Dr. Peale recommends visualizing the successful outcome of your desire or seeing it very specifically and with tremendous intensity in your mind's eye.

A wonderful example of "imaging" is Venus Williams, who won Wimbledon, the world's most prestigious tennis tournament, at the age of 20. When Williams was 10 years old, she and her father went to Florida to see the world-famous tennis champion Chris Evert. Williams put her hands on the Wimbledon trophy. From that day forward, she began to imagine herself as a champion and began to take the many steps needed to achieve her goal.

Here's a related story that most everyone in the world is familiar with: As a young child, Tiger Woods and his father began to visualize the day when he would win all the prestigious championships and be acclaimed the best golfer on the planet. His vision of where he wanted to be, plus hard work, enabled him to achieve his goal.

COMMUNITY SUPPORT GROUPS

Why are support groups needed? From the phone calls and letters I've received, I've found that people with yeast-related problems, food and chemical sensitivities and other chronic health problems like to talk to others with similar problems.

Many communities have support groups formed by people interested in nutrition, mental illness and preventive medicine. A few communities have support groups for people interested in yeast-related disorders.

Yet, patients often experience difficulty in locating a group. If that's your situation, you may want to start your own group. Here are suggestions:

- Check with the health editor of your newspaper and/or the public relations director of your hospital.
- Advertise in local papers. That way you can see if others in your area would like to join such a group.
- Do networking. Ask the staff of your local health food store or pharmacy to help you. Many stores have bulletin boards where you can post notices.
- Pass the word around to church or social groups.

See **www.yeastconnection.com** to visit an online discussion group for encouragement, ideas and success stories.

- If you know empathetic physicians, nutritionists or other professionals, ask them to help you spread the word.
- Once you get a group of 10 or 12 people, set a time and place for a meeting.
- At the meeting, select a group leader, secretary and treasurer, and organize your next meeting.
- Decide as a group the type of help/discussion you want at each meeting.
- Ask a knowledgeable professional to speak to your group.

See **www.yeastconnection.com** for a printable flyer to use as handouts, advertisements, or to post on bulletin boards.

ONLINE

There are several online support groups and more cropping up every few months. Here are a few:

- http://www.yeast.connection.com
- http://groups.msn.com/candidasupportgroup.htm.
  This is a website designed by a chiropractor and a naturopath who are involved daily in helping patients overcome yeast-related problems.
- candidiasis@yahoogroups.com.
  This public group discusses of all aspects of candida yeast overgrowth and its many implications. It is promoted by CureZone. org but is not medically supervised.

MY COMMENTS

Psychological stress can play a part in making you more susceptible to illnesses of many types, and psychological support can help you get well. In talking to my patients about this kind of support, I like to use the term "psychological vitamins." Here are some of them:

1. You need caring, empathetic people to encourage you, work with you and help you. Included, of course, would be your spouse or companion who lives with you, or a relative or best friend. It could also include a professional who understands your illness and works to help you. Support groups consisting of people who are experiencing similar problems can also help.

2. You need to be noticed, praised and encouraged. You need smiles, touching, holding, patting and petting. Physical contact stimulates the release of endorphins, a chemical that lessens anxiety and pain.

See www.yeastconnection.com to sign up for weeky e-mail help.

"Psychological vitamins" can help you regain your health and get your life back on track.

REFERENCES

1. Weiner, M.A., *Maximum Immunity,* Houghton and Mifflin, New York, 1986.

2. Galland, Leo, *Power Healing,* Random House, 1998.

3. Bock, Kenneth, M.D. and Sabin, Nellie, *The Road to Immunity: How to Survive in a Toxic World,* Pocket Books, 1997.

CHAPTER **8**

The Yeast Connection Controversy and Why It Continues

In the early and mid-1980s, millions of people learned about the relationship of *Candida albicans* to fatigue, depression and other disorders from my book, *The Yeast Connection* and Dr. Truss' book, *The Missing Diagnosis*.

This information has benefited thousands, perhaps, millions, of people, and it certainly rocked the boat in the medical establishment.

I "keep on keeping on" talking and writing about the relationship of *Candida albicans* to health problems that affect millions of people all over the world. And I rarely get discouraged because I know that people who have "rocked the boat" have been ignored, harassed or persecuted for decades and occasionally for centuries.

The Yeast Connection upset my colleagues in the American Academy of Allergy, Asthma and Immunology (AAAAI) so much that in 1986 they published a position statement, "Candidiasis Hypersensitivity Syndrome." Here are excerpts from this statement.

> This syndrome has been described and popularized by Truss and
> Crook. The syndromes are described as wide-ranging, involving mul-
> tiple systems, and include fatigue, lethargy, depression, inability to
> concentrate (and other problems).
>
> In his recommendations to patients, Crook said, "If a careful
> checkup doesn't reveal the cause of your symptoms and your medical

history (as described in this book) is typical, it's possible, even probable, that your health problems are yeast connected."

Critique: The Practice Standards Committee finds multiple problems with the candidiasis hypersensitivity syndrome.

1. The concept is speculative and unproven.
   A. The basic elements of the syndrome could apply to almost all sick patients at some time.
   B. There's no published proof that *Candida albicans* is responsible for the syndrome.
2. Elements of the proposed treatment program are potentially dangerous.

Recommendations: On the basis of evidence so far reviewed . . . the Practice Standards Committee recommends that the concept of the candidiasis hypersensitivity syndrome is unproven.

The Position Statement was signed by 11 members of the AAAAI Executive Committee.

I sent a letter of response documenting the relationship of *Candida albicans* to many chronic disorders. In spite of my efforts, this organization gave no indication that it would like additional information or to see "the other side of the coin."

I won't criticize my colleagues here, but I am writing this book so you, the medical consumer, can be the judge. For more than 20 years, I have had remarkable results with the anti-candida program, and so have many of my colleagues. It may or may not work for you, but if you have shuttled from doctor to doctor for months or years, it may be worth a try.

The medical establishment is rigidly conservative. It clings to old and time-proven ways of doing things until it is literally hit over the head with something new and, even then, most physicians and medical institutions only venture into new territory with great trepidation.

I know. I've been the subject of intense criticism and even ridicule for decades for my viewpoint on the pervasiveness of systemic *Candida albicans* infections.

(**Editor's note:** Dr. Crook didn't include here one of the prime examples of the medical community's refusal to support his theories. It seems a patient of Dr. Crook's was a young woman physician on the faculty of a major medical school who improved remarkably on his anti-yeast regimen.

The story is excerpted here from a book called *Racketeering in Medicine* by James P. Carter, M.D.

> She called Dr. Crook back and said that she wanted to organize a conference. She obtained a grant of $30,000 from Lederle Pharmaceutical Company. Her chief was not enthusiastic, but he did not oppose the holding of the conference. She invited 16 board-certified physicians, including Dr. Truss; her chief also insisted on inviting the president of the American Academy of Allergy and Immunology (sic), who would definitely be opposed to the existence of the syndrome.
>
> Six days before the conference was to take place, however, the money was withdrawn.)

I'm not alone. Scientific history is rife with the stories of pioneers who were ostracized in their time:

Galileo: This Italian astronomer and physician, born in 1564, has been called "the Founder of Modern Experimental Science." He was the first to make practical use of the telescope in studying the moon, Saturn, Venus and the Milky Way. He became a strong proponent of the Copernican theory that the sun is the center of the cosmos, with the Earth and other planets traveling around it. Although he presented his observations to Pope Paul V and other church officials, years later, at the age of 69, he published a popular book and was accused of heresy under the papacy of his friend, Urban VIII, summoned before the inquisition and sentenced to an indefinite period of imprisonment. He remained under house arrest until his death nine years later.

According to James S. Goodwin, M.D., and Michael R. Tangum, M.D., in an article published in the *Archives of Internal Medicine* in 1998 (Vol. 158, Nov. 9, 1998: 2187),

> Galileo was punished . . . for bypassing the intellectual establishment and taking his exciting ideas directly to the people . . . He was considered a usurper as well as a popularizer.

Limes and Scurvy: In the 1740s, Dr. James Lind learned by "serendipity" that putting limes and other fresh fruits and vegetables on ships of the British navy kept sailors from developing scurvy, caused by vitamin C deficiency. Yet, the "experts" didn't believe him until 45 years later when they began putting limes on all British ships, hence the slang term for British sailors, "limey." It wasn't until the 20th century that a Hungarian researcher who received the Nobel Prize discovered that limes were a valuable source of vitamin C.

Childbed Fever: In the 1840s, a 26-year-old Austrian physician, Ignace Semelweiss, gave medical students working under him the following instructions: "Wash your hands carefully before you do a pelvic examination on women in labor." Following these instructions, no more women on his obstetrical service died of "childbed fever." On other obstetrical wards in the same hospital where these precautions were not taken, 10–20% of women in labor died. When Semelweiss told his chief (an "expert" in gynecology and obstetrics) about his observations, he was fired.

Dr. Kilmer McCully: In the 1960s, this Harvard scientist described the role that elevated levels of homocysteine, a toxic product of protein metabolism, might play in causing premature death from heart disease. He also noted that adequate amounts of vitamins B_6, B_{12} and folic acid could lower homocysteine levels and heart disease risk.

In discussing Dr. McCully's observations in the June 1999 issue of his monthly newsletter, *Health and Healing,* Dr. Julian Whitaker commented:

> Thirty years ago, however, conventional medicine was so enmeshed in the cholesterol theory of heart disease that Dr. McCully's work was ignored—even ridiculed. When he refused to drop this "trivial" area of research, he was eventually asked to leave Harvard.

The observations of McCully were vindicated in a study published in the prestigious *Journal of the American Medical Association.* The Swiss authors of that article wrote:

Homocysteine-lowering therapy with folic acid, vitamin B_{12} and B_6 significantly decreases the incidence of major adverse events after . . . coronary interventions.[1]

During the last 16 years, my colleagues in AAAAI have continued to insist on placebo-controlled, double-blind studies "proving" that *Candida albicans* is responsible for health problems that affect many people. Yet, there's considerable support for the importance of clinical experience and even unblinded trials. For example, the Office of Technology Assessment (OTA), a branch of the United States Congress, in a 1978 article entitled "Assessing the Efficacy and Safety of Medical Technologies," stated:

It has been estimated that only 10–20% of all procedures currently used in medical practice have been shown to be efficacious by a controlled trial.

Since the 1985 report by allergy "leaders" was published, there have been a number of reports that support the relationship of superficial yeast infections to many chronic disorders. These include asthma, autism, chronic fatigue, endometriosis, fibromyalgia, headaches, interstitial cystitis, multiple sclerosis and psoriasis.

In 1991, Dr. Leo Galland reviewed all published studies describing the allergic and toxic effects of superficial yeast infections and substantiated these relationships.

Part of the controversy may be more easily understood, says Dr. Galland, by understanding that doctors have always recognized the candida they can see, such as vaginal and esophageal candida.

He said:

Recently doctors have become aware of another type of candida infection called disseminated candida that occurs only in severely immunocompromised individuals, such as patients on chemotherapy and drug addicts. And it is potentially life threatening. What we are talking about here with the yeast connection is a very different thing. Many doctors are still resistant to the concept of systemic yeast overgrowth that affects mainly the gut, but also affects the entire system because of allergies or a toxic mechanism.

In fact, the National Library of Medicine's PubMed directory of millions of articles published in medical journals around the world shows more than 15,000 entries mentioning the words *"Candida albicans,"* indicating it is of great interest to the medical community today.

(**Editor's note:** Nearly 240 of those entries were published in 2002 and 2003, many of them after Dr. Crook's death in October 2002. They include some interesting new research:

- a German study that shows *Candida albicans* was present in 75.5% of cystic fibrosis patients tested;[2]
- a Czech study that describes a simple new method for screening for candida species, calling the candida yeast "opportunistic pathogens associated with the rising incidence of life-threatening infections in immunocompromised individuals;"[3]
- a German study that shows *Candida albicans* overgrowth in the intestines of patients with antibiotic-resistant infections;[4]
- and a French study that shows the risk of death in critically ill patients with candida present in peritoneal (abdominal cavity) fluid.[5]

  While the medical community has long recognized systemic candida overgrowth in critically ill patients, it's only just now seeing the possibility that candida is an underlying factor in many chronic, but not life-threatening illnesses.

  Perhaps Dr. Crook's ideas are finally getting the scientific consideration he was denied in his lifetime.)

In my mind, the only three issues to consider in picking a therapy are:

1. Does it help?
2. How toxic is it?
3. How much does it cost?

The answers:

1. Absolutely yes, it helps many patients.
2. While the special diet is completely harmless and might be beneficial to almost anyone, the prescription antifungals do have

some side effects and must be monitored carefully and used according to standard dosages.

3. The special diet is inexpensive. The prescription antifungals are expensive (totaling approximately $330 a month at the outset for the most commonly used ones, which are almost always covered by medical insurance). For most patients, relief comes rather quickly (within three or four months) and maintenance doses are much lower and much more economical.

REFERENCES

1. Schnyder G. et al. Effect of homocysteine-lowering therapy with folic acid, vitamin (B12) and vitamin (B6) in clinical outcome after percutaneous coronary intervention: the Swiss Heart Study: a randomized controlled trial. JAMA 2002 Aug. 28;288(8):973–9.

2. Bakare N. et al. Prevalence of *Aspergillus fumigatus* and other fungal species in the sputum of adult patients with cystic fibrosis. Mycoses 2003 Feb;46(1–2):19–23.

3. Dostal J. et al. Simple method for screening Candida species isolates for the presence of secreted proteinases: a tool for the prediction of successful inhibitory treatment. J. Clin Microbiol 2003 Feb;41(2):712–716.

4. Krause R. et al. Elevated fecal Candida counts in patients with antibiotic associated diarrhea: role of soluble fecal substances. Clin Diag Lan Immunol 2002 Jan;10(1):167–68.

5. Dupont H. et al. Predictive factors of mortality due to polymicrobial peritonitis with candida isolation in peritoneal fluid in critically ill patients.

CHAPTER 9

Getting Help for Your Yeast-Related Problems

Finding a health professional to help with yeast-related problems can be tricky these days when insurance companies squeeze pennies and physicians every which way.

During medical school, internship and residency training, most MDs and osteopaths (DOs) were taught to recognize and treat disease. Although curriculum changes are being made in many medical schools, most MDs and DOs in practice today received little training in promoting health.

In addition to your own primary care physician, there are many caring health care professionals willing to help:

Osteopaths: DOs can perform surgery, deliver babies, treat patients and prescribe medications in hospitals and clinical settings and in all branches of the armed forces. DOs use all the tools of modern medicine, just as MDs do. In addition, DOs are trained to perform osteopathic manipulations, a technique in which a diagnosis is reached by giving special attention to the joints, bones, muscles and nerves. Manipulation improves the circulation, which, in turn, creates a normal blood supply, enabling the body to heal itself.

Naturopaths: NDs or doctors of naturopathy employ various natural means to empower an individual to achieve the highest level of health possible. NDs may use a variety of natural healing techniques, including clinical nutrition, herbal medicine, homeopathy, Oriental medicine, acupuncture, hydrotherapy, physical medicine (including massage and therapeutic manipulation), counseling and other psychotherapies and minor surgery. In several states they are licensed to write prescriptions for naturally derived drugs, including antibiotics and nystatin, one of the antifungals effective against candida yeast overgrowth.

Chiropractors: More than 50,000 DCs (doctors of chiropractic) practice in the United States and Canada. Some of these licensed health professionals restrict their practices to treatment of neuromusculoskeletal and orthopedic conditions. However, many chiropractors include clinical nutrition, dietary therapy, exercise training, lifestyle modification, environmental control and mind/body techniques. They are also trained in diagnosis and laboratory assessment and often work with other specialists. Some also have expressed an interest in helping patients with yeast-related illnesses.

Nurse Practitioners: In most states, NPs and Advanced NPs are qualified to diagnose many common and some chronic illnesses and provide long-term care working in collaboration with a physician. Nurse practitioners are usually registered nurses who have specialties in a specific field, such as gerontology, women's health, midwifery, cardiology care, etc. NPs are sometimes primary care providers for wellness exams.

Physicians' Assistants: These are health care professionals licensed to practice medicine with physician supervision. As part of their comprehensive responsibilities, PAs conduct physical exams, diagnose and treat illnesses, order and interpret tests, counsel on preventive health care, assist in surgery and, in most states write prescriptions. Some practice in rural clinics under the supervision of a physician, but the physician may not always be present.

Although NDs, NPs and PAs in many states can prescribe nystatin, you'll probably need a medical doctor to prescribe and monitor Diflucan, Nizoral, Sporanox or other systemic antifungal medications. In my opinion, these drugs are the "first team" and are needed to overcome many of the yeast-connected problems I've described in this book—along with an appropriate diet.

If your gynecologist, family physician or internist is kind and caring, even though skeptical of "the yeast connection," this still may be the best person to help you. Write your physician a letter and say something like this:

Thank you for the patience you've shown in listening to my many complaints and for the help you've given me. In spite of the tests and therapies I have received, I'm continuing to experience health problems that I feel may be yeast-related. Will you work with me?

I realize that you may not believe Candida albicans *plays a significant part in causing my symptoms. I can understand your point of view because I've read medical reports that concluded that the relationship of yeast to many chronic disorders is "speculative and unproven."*

I've also learned that no laboratory will determine absolutely whether or not my problems are candida-related. However, observations by reputable physicians I've read about in Dr. William Crook's books show that many of my symptoms can be helped if I change my diet and take Diflucan, nystatin or other antifungal medications.

Although I realize that these medications and dietary changes will not provide a "quick fix" for my problems, I feel they would be steps in the right direction.

You might want to add a diary of your medical troubles as a reminder, for example:

December 1999:   Bronchitis, given a course of antibiotics.

February 2000:   Vaginitis, given Monistat.

May 2001:   Sinus infection, given more antibiotics.

July 2001:   Two migraine attacks, given Imitrex. Migraines continue every 4–6 weeks.

November 2001:   Doctor visit for muscle and joint pain, lack of energy, constipation.

January 2002:	Doctor visit, no energy. Doctor suggests I might be depressed.
February 2002:	Bronchitis again, more antibiotics.
March 2002:	Doctor visit—bronchitis better, no energy. Told to take multivitamins.
April 2002:	Vaginitis again. More Monistat. Takes two weeks to clear up this time.
May 2002:	Doctor visit. Still no energy. Joint pain continues.

WHY AREN'T MORE MEDICAL DOCTORS INTERESTED IN TREATING PATIENTS WITH YEAST-RELATED PROBLEMS?

There are many reasons and I'll list only a few of them.

- These physicians use a number of examinations and tests that help them make a diagnosis of an illness or disorder. These include a careful history, complete physical examination, x-rays, blood tests and other laboratory tests of many types. But in spite of comprehensive examinations, no tests are available that enable the physician to make a definite diagnosis of a yeast-related problem.
- Statements by major medical organizations, including the American Academy of Allergy, Asthma and Immunology and the American Medical Association, have concluded that the pioneer observations by me and by C. Orian Truss, M.D., and the clinical observations of many physicians are "speculative and unproven."
- Although some health insurance policies will cover multiple doses of 100 or 200 mg tablets of antifungal medications for recurrent vaginal yeast infections, they will usually reject claims that contain the diagnosis of candidiasis as an explanation for a patient's fatigue, headaches, depression or other chronic symptoms.
- Many busy medical doctors, although kind and caring, work under such pressure in their practices that they're unable to devote

the time needed to take on the responsibility of the patient with yeast-related problems.

- In addition, these physicians do not have enough office staff who can spend time with their patients explaining the necessary dietary changes and other treatment measures required for women (and men and children) with yeast-related problems to regain their health.

> **In my opinion:** When a careful examination has ruled out other causes for a patient's symptoms, a sugar-free diet and a one-month trial of prescription antifungal medication is the best place to start. If you and your doctor agree you are improving and you need to continue on Diflucan for a longer period of time, your liver function must be monitored.

RESOURCES

- My website: www.yeastconnection.com, which contains information about yeast-related health problems, success stories and natural non-prescription supplements that can help you.
- The American Association of Naturopathic Physicians:
  website: www.naturopathic.org
  3201 New Mexico Ave., NW, Suite 350
  Washington, DC 20016
  Toll-free: (866) 538-2267
- American Academy of Environmental Medicine (AAEM)
  website: www.aaem.com
  7701 East Kellogg, Suite 625
  Wichita, KS 67207
  (316) 684-5500

  Physicians in this organization (including MDs and DOs) are concerned with adverse reactions experienced by individuals who have been exposed to environmental excitants. The resulting disorder, as determined by the person's susceptibility, is termed environmental illness. Individuals can be susceptible to excitants found in air, water, food, drugs and in the home, work and play environments. Many AAEM members are also knowledgeable and interested in helping people with yeast-related problems.

- American College for Advancement in Medicine (ACAM)
website: www.acam.org
23121 Verdugo Dr., Suite 204
Laguna Hills, CA 92653
(800) 532-3688
Fax: (949) 455-9679

Physicians in this society are dedicated to educating physicians on the latest findings and emerging procedures in preventive/nutritional medicine. ACAM's goals are to improve physician skills, knowledge and diagnostic procedures and to develop public awareness of alternative methods of medical treatment. Many ACAM members are also knowledgeable and interested in helping people with yeast-related problems.

- American Holistic Medical Association
(AHMA)
website: holisticmedicine.org
12101 Menaul Blvd. NE, Suite C
Albuquerque, NM 87112
(505) 292-7788
Fax: (505) 293-7582

Members of this organization (including MDs, DOs, students studying for those degrees and other licensed health practitioners) emphasize the importance of the whole person—body, mind and spirit—and the interdependence of each of these parts. AHMA members stress the importance of a cooperative relationship between practitioner and patient and encourage both parties to participate fully in health care decisions.

OTHER ORGANIZATIONS AND SUPPORT GROUPS

A number of organizations and support groups in the United States, Canada, England and other countries provide information and help for people with chronic health disorders. These include multiple chemical sensitivity syndrome (MCSS), chronic fatigue syndrome (CFS/CFIDS), fibromyalgia syndrome (FMS) and many others.

By contacting these organizations you may be able to obtain information that will help you. Many are nonprofit organizations and are staffed in part by volunteers. If you're writing, enclose a stamped, self-addressed envelope and a small donation to help cover costs.

These people and organizations have agreed to be represented here. They may be able to help you or direct you to someone in your area who can.

LINKS TO OTHER YEAST-RELATED WEBSITES

If you're like many of my patients, you'll become a voracious researcher and reader. Fortunately, there is some excellent information available on candida yeast infections and dysbiosis. These are some of the best:

- Great Plains Laboratory:
  www.greatplainslaboratory.com
- Mastering Food Allergies:
  http://www.nidlink.com/~mastent/wheatfre.html
- For gluten-free cookbooks: www.savorypalate.com
- Wisconsin Institute of Nutrition: www.nutritioninstitute.com
- The Allan Magaziner Center for Wellness and Anti-Aging Medicine: www.drmagaziner.com
- Website of Elmer M. Cranton, M.D.: www.drcranton.com
- NEEDS, INC.—Nutritional information and shopping site for the health-conscious and environmentally sensitive person: www.needs.com
- Foundation for Integrated Medicine: www.mdheal.org
- East Park Research supplements to enhance the immune and glandular systems: www.eastparkresearch.com

CHAPTER 10

Special Message to Physicians

I think it is important for you to know a little more about how and why a physician with traditional medical education and training became a writer and a "maverick" who has written about and advocated for the "yeast connection" for decades.

During my years of general pediatric practice in my hometown of Jackson, Tennessee, I attended many medical meetings, including those sponsored by the American Academy of Pediatrics. I also listened carefully to what many of my patients told me, and I learned a great deal from them. I'm not saying that I believed everything they said, but I soon learned that there was an element of truth in many of their observations.

Then, through a series of serendipitous events, I wrote a book entitled *Answering Parents Questions,* which was published by Charles Thomas in 1963. This book led Milton Levine, a New York pediatrician, to invite me to write a nationally syndicated newspaper column, "Child Care." I wrote this column six times a week for 11½ years and three times a week for another three years before giving it up in 1979.

I learned many things from people who wrote me, and I published several other books in the 1970s, including *Your Allergic Child* and *Can Your Child Read? Is He Hyperactive?*

In reviewing medical literature as early as the 1950s, I was surprised to learn that sensitivity to cow's milk and other foods could cause many children to become irritable, inattentive and depressed. I found I was able to help many of my patients with food allergies and sensitivities by eliminating most sugars from their diets.

Since my first report on this discovery in 1961, I began a crusade to educate doctors, health care workers and patients about this widely diverse group of symptoms that includes recurrent ear problems, headaches, fatigue, muscle aches and respiratory problems, as well as ADHD.

In the mid 1970s, I had a patient I couldn't help. Eventually, she moved away from Jackson, and when she came back a few years later, she was well. I asked her how she got better, and she told me about Dr. C. Orian Truss of Birmingham, Alabama, and his work with *Candida albicans*. I contacted Dr. Truss, and although I admit I was skeptical at first, I tried putting some of my chronically ill patients on the sugar-free diet and the antifungal drug nystatin, and they got better.

Another leap of awareness came in 1979, when I read an article in an obscure Canadian medical journal describing the relationship between the common yeast *Candida albicans* and health problems affecting many adults.

During the 1970s and early 1980s, I appeared several times on the Bob Braun television show on WLW-TV in Cincinnati to talk about children's problems.

In December 1982, I wrote to Bob and told him I was coming to Cincinnati in January. He asked, "Dr. Crook, do you have a topic that would interest people who do not have children?" I said, "Yes, and it has to do with 'the yeast connection.'"

He asked me what that was, I gave him a brief explanation and he invited me to do another program with him and talk about it.

At the end of this popular regional show, Bob said, "Where can my viewers obtain additional information?" Rather nonchalantly I said, "They can send me a stamped, self-addressed envelope, and I'll send them four pages of notes I prepared for my own patients."

Two days later I received 1,000 letters, the following day 1,400 letters and in one week, I had received 7,300 letters!

This response led me to put aside another book about behavior and learning problems in children and write *The Yeast Connection,* which was published in December 1983.

That book changed my life and my practice. Since then, I've written six other books dealing with candida-related health problems, including the paperback edition of *The Yeast Connection* (which has sold more than a million-and-a-half copies), *The Yeast Connection Cookbook, Chronic Fatigue Syndrome and the "Yeast Connection,"* (which was replaced with *Tired—So Tired! and the Yeast Connection* in 2001), *The Yeast Connection and the Woman, The Yeast Connection Handbook* and *Yeast Connection Success Stories.*

See **www.yeastconnection.com** to purchase other bookc by Dr. Crook.

During the last 19 years, I've knocked on countless doors in my efforts to gain credibility for the yeast/human interaction. Although I haven't had much success, I have made a few friends, including John E. Bennett, a mycologist at the National Institute of Allergy and Infectious Diseases in Bethesda, Maryland, who said:

> Few illnesses have sparked as much hostility between the medical community and a segment of the lay public as the chronic candidiasis syndrome . . . Those who argue for the existence of (this) syndrome . . . have leveled a serious charge against the medical community, claiming it is not fulfilling one of its most important obligations to its patients. This charge is simply put: You physicians are not listening to your patients.[1]

Another friend who supported me, Douglas H. Sandberg, M.D., Professor of Pediatrics at the University of Miami in Florida, said:

> Confirmation of the diagnosis remains difficult, evaluation of efficacy of therapeutic measures incomplete, and tools for monitoring a therapeutic response are below the standards we've come to expect in modern medical practice.
>
> In spite of these shortcomings, I'm convinced that this disorder exists and that it is important. It must be considered in differential diagnosis of patients with a variety of chronic complaints. Since diagnosis at times can be made only through de-

termining response to a therapeutic trial, some patients would have to be treated without a firm diagnosis prior to institution of therapy.[2]

Another friend, James H. Brodsky, M.D., a diplomate of the American Board of Internal Medicine, a member of the American College of Physicians and a member of the clinical faculty of Georgetown University Medical School in Washington, D.C., commented:

> Since my introduction to the relationship between yeast and human illness in the early 1980s, I've seen well over 1,000 patients with some form of yeast-related illness . . . I maintain a general internal medicine practice and make hospital rounds daily. While I find all aspects of my practice fulfilling, nothing has been so rewarding as helping patients with yeast-related illnesses who have been unable to find help elsewhere.[3]

You may have read early papers suggesting the relationship of *Candida albicans* infections to a number of chronic health disorders is "speculative and unproven." Those papers are based on flawed scientific reasoning. The primary caveat is that these conditions cannot be treated by prescription medications alone. The low-sugar dietary component is as least as important as the pharmacological component of the treatment plan.

CLINICAL REPORTS OF THE EFFECTIVENESS OF A THERAPY OFTEN PRECEDE SCIENTIFIC STUDIES

Clinical reports that describe the effectiveness of a particular method of therapy may precede the scientific studies that provide support for the therapy by decades (or even centuries).

Sadly, there have been very few clinical studies in this field, partly because of the enormous expense of such research and partly because of a frank lack of interest from the medical and scientific communities.

Nineteen years ago, in an article published in the *Journal of the American Medical Association,* two physicians from the University of

New Mexico School of Medicine pointed out that an effective treatment for a particular disease is often ignored or rejected because the reasons the therapy worked aren't understood. And they said that the only three issues that matter in choosing a therapy are: Does it help? How toxic is it? How much does it cost?[4]

Several thousand physicians in practice and a handful of academicians have found that a sugar- and yeast-free special diet and nystatin, ketoconazole (Nizoral), fluconazole (Diflucan) or itraconazole (Sporanox) are effective in treating patients with a diverse group of seemingly unrelated health problems. These range from PMS, chronic fatigue syndrome, vaginitis, migraines, fibromyalgia, irritable bowel syndrome, interstitial cystitis and psoriasis in adults to recurrent ear infections (and respiratory infections) and the subsequent development of hyperactivity, attention deficits and autism in children.

I hope you'll take a careful look at the relationship of apparently superficial yeast infections to chronic health disorders that affect people of all ages and both sexes. Included especially are perimenopausal women who feel "sick all over" and other family members who give a history of repeated courses of broad-spectrum antibiotic drugs. I feel that in so doing, you'll be able to help many of your difficult patients and, at the same time, make your own practice more interesting and rewarding.

REFERENCES

1. Bennett, J. E., "Searching for the Yeast Connection," *N. Engl. J. Med.*, 1990; 323:1766–1767.

2. Sandberg, D. H., Statement, "Candida-Related Illness," September 22, 1989.

3. Brodsky, J. H., Statement, "The Importance of Candida-Related Health Problems," October 14, 1993.

4. Goodwin, J. S. and Goodwin, J. M., "The Tomato Effect," *JAMA*, 1984, 251:2287–2290.

PART 2

Yeast-Related Problems That Affect Women

CHAPTER 11

Premenstrual Syndrome (PMS)

What is PMS? It's a conglomeration of symptoms that affect women, primarily during the week before their menstrual periods.

Several times each week, the average American man or woman hears and reads about PMS. Frequent media attention has now made PMS a household word. I even ran across a cartoon showing one moppet whispering to his playmate (with an irate mother in the background):

"Don't worry, Mama's PMS is making her grouchy."

PMS is "for real." No doubt about it. Because I'm the father of three daughters and because most of the patients I saw during the past decade are women, I'm especially interested in the subject.

You've probably experienced it yourself, since 90% of all women deal with the aches, pains and emotional stress of PMS at some point in their lifetimes. More than a third of all women struggle with PMS for years. National health statistics show that 30–40% of women have symptoms serious enough to interfere with their everyday lives, and a smaller percentage are incapacitated by PMS from time to time.

SYMPTOMS

Here's a grim laundry list of symptoms:

- abdominal bloating
- acne

- anxiety
- backache
- breast swelling and tenderness
- cramps
- depression
- food cravings
- fainting spells
- fatigue
- headaches
- insomnia
- altered sex drive
- swelling of fingers and ankles
- personality changes including:
  —mood swings
  —outbursts of anger
  —thoughts of suicide

See www.yeastconnection.com for a printable monthly chart to help you track monthly patterns in your PMS symptoms.

CAUSES OF PMS

PMS begins during the luteal phase of menstruation, the phase right after ovulation, which can last as long as 14 days before the menstrual period begins. It usually disappears as soon as the menstrual flow ends.

It can be caused by stress, genetics, age, the number of children a woman has had, alcohol, sugar and caffeine intake, other dietary factors, lack of exercise, hypothyroidism and depression.

There is scientific evidence that women who exercise regularly are

See www.yeastconnection.com for a monthly flow chart of hormone flunctuations which impact PMS symptoms.

less affected by mood swings and depression and have fewer of the other annoying symptoms of PMS than sedentary women.

TREATMENT

Proper nutrition is also an important element in beating PMS. Not only do high sugar, alcohol and caffeine consumption practically guarantee you will suffer from PMS, general good nutrition is one of the most important factors. One study showed women most likely to suffer from PMS eat 62% more refined carbohydrates, 75% more refined sugar, 79% more dairy products, 78% more sodium, 53% less iron, 77% less manganese and 53% less zinc than those who practiced the Standard American Diet. Research has shown many women with severe PMS are likely to be deficient in B vitamins and magnesium. And a fascinating study from Columbia University showed that 1,200 mg of calcium supplementation for three months relieved 50% of the symptoms of PMS.

There is also evidence that evening primrose oil, taken with flaxseed oil or fish oil for a full complement of essential fatty acids, will help relieve symptoms.

Mini doses of natural progesterone (not progestin) have also proven helpful in relieving PMS symptoms for some women.

THE YEAST CONNECTION

PMS is frequently included among the myriad troubles women with candida yeast-related problems experience.

If the sugar and refined carbohydrate figures above are ringing any bells with you, you're right: There is an element of candida yeast overgrowth in many women with PMS.

Some researchers suggest the candida may trigger PMS symptoms by activating an autoimmune response to sex hormones such as estrogen. The cyclic rise and fall of these hormones could be an explanation for the flare-ups and then calming down of candida symptoms.[1]

In my own practice during the 1980s, I saw dozens of women with yeast-related health problems. Most of these patients experienced fatigue, headaches, irritability, bloating and depression—especially the week before their periods.

I found that an antifungal therapy and the anti-yeast diet helped many of them.

PHYSICIANS' COMMENTS

Richard Mabray, M.D., a Victoria, Texas, board-certified gynecologist, tells me he has had good results with progesterone, given as intramuscular shots and sublingual (under the tongue) drops.

Other doctors give women sublingual drops as needed throughout their menstrual cycles and increase the amount they take during the days when they most frequently experience PMS symptoms.

Another colleague, Russell Roby, M.D., of Austin, Texas, used this method. He responded to my questions about PMS and candida:

> When I see a woman with recurrent PMS, I look at thyroid first, insulin metabolism and diet second, and then I start looking at allergies and candida as a third choice. . . . In answer to your question about whether I've found anti-yeast treatment helps other symptoms—my answer is "Absolutely yes." Because I tend to "shotgun" my treatment program for women with these problems, I don't know which therapeutic intervention helps the most. I just know that it's a combination, and if I leave anything out, they're not as likely to get well.

I'm also impressed by the success rate of another colleague, Pamela Morford, M.D., of Tucson, Arizona.

> I've probably treated 400 to 500 women with PMS. The majority came in complaining of bloating, irritability and depression before their periods. Many were very concerned about being out of control and unable to handle their anger. Some had lost their confidence. Another major complaint: "spaciness' and inability to concentrate.
>
> Within a month after starting anti-yeast treatment, includ-

ing Diflucan (200 mg. for several weeks), many of these symptoms diminish considerably and occasionally disappear. Many women, however, found that they could not vary much from their diet without experiencing a return of their symptoms.

Treating PMS patients with antifungal medications is standard practice at the PMS clinic at St. Luke's Hospital in Denver.

Nystatin therapy and diet are the primary means of treating 90% of the patients who come to the clinic, says Jean Rowe, R.N., the clinic's director. She also uses low-dose estrogen and progesterone therapy for some patients. But candida is a prime consideration:

I would say that most every woman I see who has PMS has a lot of symptoms that resemble yeast overload. I tell them if they'll treat the yeast, it will give them a lot of benefit besides helping their PMS. So they get healthier and sometimes slimmer, which they really like.

Dr. Jay Schinfeld published a controlled study indicating that nystatin relieves PMS in women with recurrent vaginal infections.

NATURAL APPROACHES

There are dozens of natural approaches to PMS, including some discussed above.

One often recommended herbal remedy is chasteberry or *Vitex agnus castus*. This herb has been used traditionally to relieve menstrual disorders. It contains mild levels of hormones such as androstenedione and progesterone. Doses up to 500 mg. a day are considered safe.

In an article published in a 2001 issue of the *British Medical Journal,* chasteberry was found to cause "significant improvement" for sufferers of PMS.[2]

The progesteronic action attributed to chasteberry seems to be important in relieving symptoms of PMS. Many practitioners recommend getting a topical cream containing chasteberry. Chasteberry is an ingredient in some progesterone creams, including Phyto-Gest and Estro-All.

Naturopathic doctor Tori Hudson, N.D., author of *Women's Encyclopedia of Natural Medicine,* adds, "Since natural progesterone is a hormone, I think it is best to seek the advice of a qualified health care practitioner who is experienced with its use. This assures proper usage and therefore maximum results."

Other natural recommendations come from Michael Murray, N.D., who offers a laundry list of lifestyle recommendations:

- Follow a vegetarian or predominantly vegetarian diet.
- Reduce your intake of fat.
- Eliminate sugar from your diet.
- Reduce your exposure to environmental estrogens.
- Increase your intake of soy foods.
- Eliminate caffeine from your diet.
- Keep your salt intake low.
  Supplement your diet with vitamins and minerals.
- Select appropriate herbal support. (He recommends chasteberry in tablet or capsule form, 175–225 mg. daily.)

And, he says, if you're bothered by PMS water retention, take licorice root three times a day, beginning on the 14th day of your cycle, and continue until menstruation begins.

Dr. Dean wrote about the natural treatment for PMS in several journal articles[3,4] and in her practice had great success with the following nutrients taken twice daily two weeks before the beginning of your menstrual period:

- Vitamin B6, 50–100 mg.
- Magnesium, 400 mg.
- Gamma linoleic acid, 120–240 mg., in the form of evening primrose oil, borage, or black current oil

Good vitamin and mineral supplementation is important to overall good health, as is addressing the deficiency of B vitamins in women with PMS. One scientific study affirms the need for vitamin B_6. The author of the study published in the *British Medical Journal* wrote: "Results suggest that doses of vitamin B_6 up to 100 mg/day are likely to be of benefit in treating premenstrual symptoms and premenstrual depression."[5]

Other natural ways of approaching PMS:

1. deep breathing for 5–10 minutes twice a day
2. exercise
3. aromatherapy using essential oils such as sandalwood, juniper or geranium

MY COMMENTS

PMS, like all of the disorders I've discussed in this book, develops from a combination of many different causes. There is rarely a quick fix. But if you're bothered by PMS and made a high score on the yeast questionnaire, a sugar-free special diet, probiotics, oral antifungal medications and nutritional supplements could change your life.

RESOURCES

Murray, Michael T., Pizzorno, Joseph E., *Encyclopedia of Natural Healing,* Prima Publishing, 1998.

REFERENCES

1. Severino, S. K., Moline Ml. "Premenstrual syndrome: clinician's guide." New York: Guilford Press, 1989;130.

2. R. Schellenberg, "Treatment for the premenstrual syndrome with agnus castus fruit extract," *British Medical Journal,* 2001 (January 20);322(7279):134–7.

3. Dean, C., et al, "Medical Management of Premenstrual Tension," *Canadian Family Physician,* February 1986.

4. Dean, C., et al, "Pharmacology of Premenstrual Tension," *Canadian Journal of Pharmacology,* February 1985.

5. Wyatt, K. M., "Efficacy of vitamin B-6 in the treatment of premenstrual syndrome: systematic review," *British Medical Journal,* 1999 (May 22);318(7195): 1375–81.

CHAPTER 12

Vaginitis

In the past 10 years or so, an area of women's anatomy rarely mentioned in public has become front-page news: the vagina.

A major factor in prompting this publicity was the decision by the medical establishment powers that be to allow women to purchase anti-yeast vaginal suppositories over the counter.

Unless you're blind, deaf, dumb and never turn on your TV or look at a magazine, you can't have escaped advertisements touting Monistat 7, Gyne-Lotrimin and other anti-yeast vaginal suppositories for yeast infections.

Vaginal yeast infections are so common that statistics show virtually every woman is likely to experience them at least once in her lifetime. One colleague of mine who runs a women's health practice says that 50–60% of her patients come to her with some form of vaginitis. Unfortunately, many women are plagued by infection after infection. The truth is that many of them had one episode of candida yeast infection that was never adequately addressed, so the yeast ran rampant in their vaginas and eventually throughout their systems.

It's important to note here that not all vaginal infections are caused by *Candida albicans*. Similar symptoms can be caused by bacterial vaginal infections, parasitic infections, sexually transmitted diseases, allergies and even hormonal changes. There are also other candida yeast infections caused by organisms other than the *Candida albicans* species. The difference is only evident with microscopic examination.

Although most professionals agree that the *Candida albicans* yeast is the most common cause of vaginitis, it's advisable for a woman to go to a health professional for a careful examination before self-treating a vaginal problem. The physician should examine vaginal discharges or secretions under a microscope to determine the cause or causes of the vaginitis so that an appropriate treatment program can be prescribed and recommended. Unfortunately, that is rarely done.

Most cases of vaginitis are characterized by itching and burning in the vagina and an unusual discharge, sometimes with a strong odor. However, many women experience no symptoms whatsoever.

If you think you may have vaginitis, check with your doctor before trying over-the-counter medications. While they may work just fine, your doctor will want to monitor your progress and try something else if you don't get complete relief in three days.

THOUGHTS FROM OTHER PHYSICIANS

Donald Lewis, M.D., my recently retired colleague here in Jackson, Tennessee, says, "If a regular patient calls up and talks to me (or to the nurse practitioner who works with me) and says she thinks she has a vaginal yeast infection, and she hasn't had one in five years, and asks me if it's okay to take an over-the-counter preparation, I'll usually tell her it's OK.

"But I also tell her that if the vaginal symptoms do not clear up in three days, then she should come in and let me do a careful examination. The nature of the vaginal discharge is important, including whether or not it has an odor. I also make a microscopic examination of the bacterial smear and determine the pH."

Another colleague, Daron G. Ferris, M.D., a professor of obstetrics and gynecology and family medicine at the Medical College of Georgia with a special interest in treating vaginitis, cautions women against self-diagnosing a vaginal infection.

Dr. Ferris found that about 65% of women who diagnosed themselves with a vaginal yeast infection were actually incorrect in their diagnosis and, by taking over-the-counter antifungal medications designed to eradicate candida yeast, they actually wasted their money and prolonged their discomfort until a correct diagnosis could be made.[1]

RECURRENT VAGINITIS

Here's why I think so many women are troubled by recurrent vaginal yeast infections:

- Candida yeasts thrive in a dark, warm, moist environment. The vagina and the intestinal tract serve as ideal places for these yeasts to live and multiply—especially when other factors encourage their growth.
- Broad-spectrum antibiotics. These drugs are often prescribed for children with ear infections and respiratory problems, teenagers with acne and women with cystitis, even for people with colds and flu, against which they have no effect. Many of these antibiotics are prescribed needlessly. While eradicating bacterial "enemies," they also knock out bacterial "friends," including *Lactobacillus acidophilus* and other naturally occurring friendly bacteria. They do nothing whatsoever to kill viruses that cause colds and flu. As a result, the usually benign *Candida albicans* multiplies and causes infection. Antibiotic-laden chickens and antibiotics in other animal products we eat also disturb the normal balance of bacteria in the intestinal tract.

If you've had four or more vaginal yeast infections in a year, you have recurrent vaginitis.

Broad-spectrum antibiotics kill the "good" and the "bad" bacteria indiscriminately. When the bacterial balance gets out of whack, yeast grows unchecked.

There is a connection between yeast in the digestive tract and vaginal yeast, according to a 1977 study carried out by Mary Ryan Miles, M.D., and two of her colleagues at Michigan State University on 98 young women who were troubled by recurrent vaginitis. Even though the study was conducted 25 years ago, the basic finding is still valid: All of these women showed yeast in the stool. Here's an excerpt from the abstract of this study published in *The Journal of the American Medical Association.*

Ninety-eight young women who complained of recurrent vaginitis were selected in sequence. The results showed that if *C. albicans* was cultured from the vagina, it was always found in the stool.

Conversely, if it was not isolated from the stool, it was never found in the vagina. These data are presented as an explanation for the recurrent nature of candida vaginitis *and thus a cure for vaginitis would not be possible without prior eradication of C. albicans from the gut . . .*[2]

Many physicians prescribe oral nystatin for recurrent vaginal yeast infections. You may need to take it as much as four times a day for as long as three months. You may be able to taper off your dosage after you've been on it between one and three months.

Others prefer Diflucan for recurrent yeast-related vaginitis. Some patients will need to take it for several months. Diflucan is also commonly prescribed for an occasional bout with vaginitis.

Some physicians, like my colleague George Kroker, M.D., of La-Crosse, Wisconsin, have very different viewpoints on vaginitis. "In my opinion, the vagina, like the nose and sinuses, is an allergic organ," says Dr. Kroker. Over the years, he says he has noticed that many women complain of intense vaginal itching and burning, and they notice the same symptoms when they are exposed to airborne molds or allergenic foods.

Other physicians have suggested women may develop vaginal symptoms due to allergies to semen or latex like that found in condoms.

"I wish that most physicians, particularly gynecologists, would think of the vaginal lining as an active immunologic organ rather than a reproductive receptacle occasionally prone to infection," Dr. Kroker concludes.

BACTERIAL VAGINOSIS (BV)

Bacterial vaginosis is caused by overgrowth of bacteria normally found in the vaginal area. The main symptom is an increase in vaginal discharge. While normal vaginal discharge is a milky white, discharge with this condition may be grayish white or yellow and be thin and watery with a strong fishy odor that is sometimes more noticeable after intercourse.

BV is usually treated with antibiotics, which then can give rise to yeast overgrowth and a vicious cycle of infection, and then the emergence of a different infection.

Experts in the field warn that physicians should be concerned about the increase in the incidence of BV, which, in the past, was considered a nuisance condition and was treated to eliminate annoying symptoms. However, recent studies showed that BV may be associated with "potentially serious infections, including upper genital tract infections, pregnancy complications and an increased risk of acquiring HIV and other sexually transmitted diseases," according to the National Vaginitis Association's 2002 report.

Among the recommendations in that report:

- Physicians should encourage women to talk about their vaginal health.
- Physicians should discuss with you what is normal. Many women think a fishy odor is normal. It is not.
- Women should not use chemical douches and perfumed sanitary pads.

NATURAL MEDICINE

During the mid and late 1990s, I became increasingly interested in natural medicine. I read a number of books about nonprescription remedies people could use for common complaints.

During the past two decades, I have attended many regional and national health food conventions and been a guest speaker at many of them. On several occasions at these conventions, I've visited with Dr. Michael Murray, N.D., one of the world's leading authorities on natural medicine. He's a graduate, faculty member and member of the Board of Trustees of Bastyr University in Seattle.

In addition to maintaining a private medical practice, Dr. Murray has written numerous books and has an interest in candida yeast infections. In the foreword to his book, *Chronic Candidiasis—The Yeast Syndrome,* he says:

> One of the great myths about natural medicines is that it is not scientific. The fact of the matter is that for most common illnesses, there is greater support in the medical literature for a natural approach than there is for drugs and surgery . . . *Unfortunately for many people, they are never aware of the natural approach that can put them on the road to lifetime health.*

Dr. Murray suggests that successful treatment for yeast infections involves re-establishing the normal bacteria found in the vagina by douching with acidophilus twice a day plus taking an oral probiotic supplement (see Chapter 30) to re-establish balance as quickly as possible.

To make an acidophilus douche, add two teaspoons of acidophilus powder to 1 quart of warm water. Douche once or twice a day. Do not use an acidophilus douche for more than five days because long-term use can cause irritation of the vaginal walls.

Dr. Tori Hudson, N.D., a highly respected proponent of natural health and professor of gynecology at the National College of Naturopathic Medicine in Portland, Oregon, reminds us to look at vaginitis holistically and systemically.

"The health of the entire body affects the vagina and the body's health depends on many things," she says.

In addition to a low-sugar diet, Dr. Hudson recommends vitamin E, vitamin A and beta carotene and a number of botanical medicines, including garlic, golden seal, Oregon grape root and tea (or ti) tree oil, prepared in combination in a douche or a suppository.

"Accurate diagnosis is the most important key to efficient appropriate treatment, whether the therapies are natural or pharmaceutical. If you know what kind of infection you currently have and choose self-treatment, it is essential to recognize when and if self-treatment isn't working and to seek professional care at that time," writes Hudson.

RECOMMENDATIONS TO PREVENT FUTURE OUTBREAKS OF CANDIDA YEAST INFECTIONS

1. Eliminate simple sugars from your diet. I can't emphasize this enough. Animal studies have shown that rats fed sugar water had 200 times the intestinal growth of candida as those who got plain water.
2. Avoid alcoholic beverages.
3. Eat some plain, unsweetened organic yogurt to help re-establish friendly bacterial colonies in the intestines, thereby balancing the yeast in the gut. We recommend Stonyfield Farm yogurt.
4. Wear panties with a cotton crotch.
5. For some women, a 600 mg. boric acid suppository at bedtime can help change the acidity of the vagina, making it a less hospitable home for candida and bacteria.

MY COMMENTS

If you're bothered by recurrent vaginal problems, in spite of examinations and treatment by competent physicians and/or other health professionals, "take charge" and learn everything you can about the many things that are important—even essential—if you want to regain your health.

Read, read and read some more. Look at the "Resources" sections of the individual chapters in this book. Read articles in magazines like *Delicious!, Let's Live, Better Nutrition, Natural Health, Prevention, Consumers' Research* and *Bottom Line Health.*

Read other books you'll find in your health food store, in the women's section of general bookstores and in your public library; and search the Internet. Become an authority and tell your physician or other health professional about what you've learned.

As you and I know, every person differs from every other person, and as you already know, there's no "magic cure" or "quick fix" for recurrent vaginitis—or any other health problem.

But the more you know, and the more you do, the greater your chances will be of not only conquering your vaginitis, but also overcoming symptoms discussed elsewhere in this book.

RESOURCES

The National Vaginitis Association's website, www.vaginalinfections. com, provides vaginal health resources for patients and health care practitioners, including the consumer brochure, "Women's Guide to Vaginal Infections."

REFERENCES

1. Ferris, D. G., Nyirjesy, P., et al. "Over-the-counter antifungal drug misuse associated with patient-diagnosed vulvovaginal candidiasis," *Obstetrics and Gynecology,* 2002 Mar;99(3):419–25.

2. Miles, M. R., Olsen, L. and Rogers, A., "Recurrent Vaginal Candidiasis: Importance of an Intestinal Reservoir," *JAMA,* 238:1836–1837, Oct. 28, 1977.

CHAPTER 13

Endometriosis

Endometriosis is a painful chronic condition that affects 5.5 million women in North America. It occurs when tissue similar to the endometrium, or the lining of the uterus, is found in locations outside the uterus. It may be found on ovaries, the outside of the uterus, the bowel, bladder, the ligaments that hold the bladder in place or the peritoneum (the lining of the pelvis and abdominal cavity). In rare cases, it can be found in far distant sites, such as scar tissue on an arm or leg.

The misplaced tissue develops into growths that respond to the menstrual cycle the same way the tissue of the lining of the uterus does. Each month, the tissue builds up and sheds. While menstrual blood flows out of the body from the uterus through the cervix and vagina, the endometriosis tissue and cells it sheds have no way of leaving the body, so they can cause inflammation, scar tissue formation, adhesions and bowel problems.

SYMPTOMS

The Endometriosis Association's website (www.endometriosisassn. org) cites the following symptoms of endometriosis:

- pain before and during periods
- pain with sex
- infertility
- fatigue
- painful urination during periods
- painful bowel movements during periods

- other gastrointestinal upsets such as diarrhea, constipation and nausea

In addition, the Endometriosis Association (EA) says it is now becoming apparent that women with endometriosis are also more apt to be troubled by:

- chemical sensitivities
- chronic fatigue syndrome (CFS)
- asthma and eczema
- infections
- food intolerances
- mononucleosis
- mitral valve prolapse
- fibromyalgia
- autoimmune disorders, including lupus and Hashimoto's thyroiditis

CAUSE

While no one knows the exact cause of endometriosis, the most popular theory is called "retrograde menstruation," in which endometrial cells from the uterus somehow travel through the fallopian tubes and into the abdominal cavity. That theory is currently being questioned. But there are other theories that indicate there may be genetic predispositions to endometriosis, or that it can be caused by an immune system malfunction and even by environmental toxins, especially dioxin.

In fact, one animal study shows that 79% of monkeys exposed to dioxin in their food developed endometriosis within 10 years.[1]

Of course, the possibility of environmental influences will probably make you perk up your ears, since we're talking here about yeast overgrowth and we already know there is a remarkable connection between environmental toxins, food sensitivities and yeast overgrowth.

In addition, says the EA, many women with endometriosis suffer from allergies, chemical sensitivities and frequent yeast infections.

YEAST CONNECTION

In many ways, the discovery of the depth of the candida problem has been a challenge for me, much like unraveling a mystery. Each part seems to lead to another part.

And so endometriosis and candida yeast infections have a connection, at least some of the time.

In the early 1980s, John Curlin, M.D., a Tennessee gynecologist, began using nystatin to treat some of his patients with menstrual irregularities, pelvic pain and endometriosis. His prognosis: "Although this program doesn't relieve all of the symptoms in women with these problems, the response in many of my patients has been gratifying."

I'm not alone in my suspicion that there is a strong link between candidiasis and endometriosis. The Endometriosis Association's book, *Overcoming Endometriosis,* has an entire chapter on the connection with candida.

It quotes Dr. Truss as follows:

I think it is unquestionable that there's a very high association of endometriosis with chronic candidiasis. Naturally, we cannot at this time be sure whether the yeast is causing the endometriosis, or whether some common factor predisposes to both.

Many doctors have shared this theory with me.

Tucson, Arizona gynecologist Pamela Morford, has another interesting viewpoint on the connection between candida and allergies in women with endometriosis.

She says, "I've found that people with endometriosis and pelvic pain frequently have a problem with yeast and allergies. And if I treat a patient with endometriosis using anti-candida therapy, she will more than likely lose her pelvic pain . . . I won't say that the endometriosis goes away, yet I will say that the pain and the symptoms go away . . . Endometriosis is an autoimmune disease of the pelvis and in those people who are genetically predisposed to developing it, candida may interfere with the immune system functioning and allow the disease to become manifest. At least, that's my theory."

Dr. Wayne Konetzki, M.D., a Waukesha, Wisconsin allergist, has done some research of his own on the immune system problem that suggests some women may be allergic to their own hormones. In a study of his patients, he found that women with endometriosis are most likely to be allergic or sensitive to luteinizing hormone and estrogen, as well as *Candida albicans,* chemicals and foods, but any combination of sensitivities is possible. Interestingly, he found that women with PMS were most likely to be allergic or sensitive to progesterone.

Other scientists have confirmed the idea that a woman's immune system, in building antibodies to *Candida albicans,* may be building antibodies against her own ovarian tissue.

AN EXPERT'S OPINION

Sometimes, in fact, many times, dedicated, caring, persistent people can advance the frontiers of medicine more than physicians and other "experts." Mary Lou Ballweg, who co-founded the Endometriosis Association (EA) with Carolyn Keith in 1980, is one of those people. Herself a victim of what she calls "this painful, chronic and stubborn disease," Ballweg has devoted her tireless efforts to informing women and health care professionals about endometriosis and offering support to the millions of women who are suffering from the disease.

"Endometriosis is a disease affecting an estimated 5 million women in the U.S. and millions more worldwide. It is a nightmare of misinformation, myths, taboos, lack of diagnosis and problematic hit-and-miss treatments overlaid on a painful, chronic, stubborn disease," says Ballweg, who serves as the Association's president.

"Women with this disease have been much maligned—supposedly they were white, stressed out, perfectionistic, upper-socioeconomic-level women who brought the disease on themselves by postponing childbearing. Only when the Endometriosis Association began in 1980 and systematically gathered data, were we able to disprove all of these myths . . . Endometriosis is, in fact, an equal opportunity disease affecting all races, personalities, socioeconomic groups, as well as all ages of females, from as young as 9 to as old as women in their 60s and 70s."

Ballweg has a long-held interest in the connection between candida and endometriosis. "I'm not sure it's always vaginal yeast infections. Instead, it may be yeast overgrowth in the intestinal tract," she says.

Ballweg thinks the entire anti-yeast treatment program, including antifungal medications and the sugar-free diet, have helped many EA members overcome their symptoms. "The diet is very important. If a woman just takes the anti-yeast drugs, that doesn't do it," she says.

Some previously closed minds are beginning to open-just a bit. Ballweg told me the story of the EA's 20th anniversary conference in 2000:

> We had some of the best people in the world there. . . It was very exhilarating for me because it is clear that the leaders in our field think that endometriosis is a systemic disease. They agreed on that. Even five years ago that would have been impossible.

Ballweg had another piece of big news in 2002 in a collaborative study between the EA and the National Institutes of Health:

> And with all their resources, we were able to run thousands of cross tabulations in our data, and all of the statistics clearly show that women with endometriosis have a greater risk for autoimmune diseases.

TREATMENT

Sometimes the best a woman with endometriosis can expect from mainstream medicine is over-the-counter pain medication—aspirin, acetaminophen, ibuprofen, naproxen and, in some cases, prescription pain medications.

Sometimes hormone therapy is prescribed, including birth control pills, progesterone (natural progesterone, not progestin), a testosterone derivative and other types of hormones. Side effects from these drugs can be problematic for some people. In her practice, Dr. Dean found that endometriosis and PMS respond particularly well to natural progesterone. She recommends being sure your doctor understands that synthetic progestin will not work the same way and may have harmful effects.

Eventually, as the lesions enlarge, surgery often becomes the treatment of choice. It can be laparoscopic (belly button) surgery, where misplaced tissue can be removed, or radical surgery with a full incision that often includes a hysterectomy, the removal of all growths and sometimes the ovaries.

Many physicians now discourage surgery, which removes the lesions and growths, but leaves the underlying cause intact.

Knowing what we now know about candida yeast overgrowth, if this is the cause of your endometriosis, none of these treatment options is likely to resolve the problem.

Dr. Konetzki says endometriosis is an extremely complex disease:

One of the most important concepts to grasp initially is that endometriosis is a systemic disease and not a local problem.

Careful history-taking reveals that patients very often have not only pain but also symptoms of nervousness, tension, anxiety, headaches, depression, fatigue, insomnia, food cravings, muscle and joint pains, indigestion, gas, bloating, diarrhea and/or constipation, recurrent vaginitis, recurrent cystitis, interstitial cystitis and loss of libido.

In his continuing discussion, Dr. Konetzki says that menstrual cramps, PMS and endometriosis are closely related and caused by a woman's hypersensitivity to her own hormones.

Dr. Konetzki says that candida plays an important role in causing symptoms in almost all of his endometriosis patients and that treatment of these patients has three components:

1. Allergy desensitization, often through a process called oral tolerization. Patients are given daily sublingual drops that contain a small amount of the antigen thought to contribute to the disease by producing an allergic reaction in the immune system. The goal of toleraization is to eliminate the allergic reaction by building up the patient's resistance.
2. A 100% sugar-free diet that also restricts yeast-related products. And he said, "After 3–6 months of treatment, most patients are able to add yeast foods slowly back into the diet. Others need to

avoid yeast for a longer period of time. The intake of refined sugars and alcoholic beverages should be completely avoided for at least a year and then restricted."

3. Systemic prescription antifungal drugs (Nizoral, Diflucan or Sporanox) during the first 21 days followed by nystatin for at least 9 months, sometimes longer. As an alternative to nystatin, he suggests two nonprescription products available from health food stores—caprylic acid and/or grapefruit seed extract.

I was pleased, even thrilled to hear the comments of a Houston, Texas gastroenterologist, John R. Mathias, about the overall importance of diet, and especially limiting sugar intake, at the EA's 20th anniversary meeting.

Dr. Mathias cited the dramatic increase in our per capita consumption of refined sugar in the past 130 years as an explanation for the health problems we are experiencing:

> The hormone that regulates glucose transport is insulin, and insulin is the most diverse hormone in our body. Insulin has short-term functions that last from seconds to minutes, such as glucose transport; intermediate functions that last from minutes to hours; and long-term functions that last from hours to days. We have allowed insulin to become the most powerful hormone in our body because of the striking increase in the consumption of carbohydrates that have a high glycemic index such as foods made from the flour of wheat, corn and rice.
>
> We have suggested that endometriosis is a disease of insulin resistance. However, in our most recent studies, we show that all subjects with endometriosis were insulin-sensitive and developed reactive hypoglycemia (low blood sugar) to glucose challenge (when given a large dose of sugar).

Because of my intense interest in "the yeast connection," I felt that Dr. Mathias' observations explain why people with candida-related disorders "crave" sweets and why restricting intake of sugar and other simple carbohydrates is an essential part of the treatment program for all candida-related disorders and diseases.

MY COMMENTS

There are many causes of endometriosis and other women's health problems that I discuss in this book. Yet, based on reports I've received from physicians and other professionals, and from women who have written me, a treatment program for endometriosis should sharply restrict sugar. It should also include:

- a wide variety of nutritious foods
- prescription and nonprescription antifungal agents
- nutritional supplements
- the control or avoidance of chemical pollutants and molds in homes and workplaces

A comprehensive anti-candida treatment program can play a part in helping many women with endometriosis and other disorders regain their health and their lives.

CASE HISTORIES

In the first edition of this book, I included the story of Mindy, who told of her severe endometriosis. It caused her debilitating pain, exhaustion, joint pains, insomnia, and difficulty in concentrating. She improved on a comprehensive program that included nystatin, vitamins, minerals and probiotics. And she said:

> If I'm under stress or get sick, my symptoms recur, but generally I'm functioning pretty well. Whether what I have is an allergy or part of chronic fatigue syndrome, I don't know. All I know is that I have improved.

In this edition, I'm including the stories of Judy and Diana, which were originally published in the EA Newsletter.

Judy, Michigan

Dear Endometriosis Association, I was astonished to read the article on the "yeast connection." It was like reading my own life history . . .

First of all, I have been treated for severe acne since the age of 15 with tetracycline. I have been on it almost continually in low doses for 17 years. I have also been on prednisone for short periods several times for inflamed acne cysts. I was on birth control pills for about three years in my early 20s.

During this period, I never had any symptoms of vaginal yeast infections, but I remember almost every time I had a routine pelvic examination, I was told I had a yeast infection, *even though I never had any symptoms*. I frequently had rectal itching from the tetracycline, and I also had an oral yeast infection. In the past few years, I have had several yeast infections with the symptoms of itching associated with them.

I was diagnosed as having minimal endometriosis by laparoscopy at the age of 27, although I had had painful menstruation since the age of 15. During the following three years, I had extremely painful menstrual periods and began having pain at other times. Two years ago, at the age of 30, I had a hysterectomy for extensive endometriosis. I still have one ovary which occasionally hurts, but my pain is greatly relieved . . .

I've also found it very interesting to note that I've had many of the other symptoms you describe, which may be yeast-related. I have chronic nasal congestion and conjunctivitis throughout the year. I am sensitive to molds and perfumes. Testing showed I am allergic to molds. I've had various muscle aches off and on for at least 10 years. I have also had some strange neurological symptoms (vision problems, dizziness and poor balance, weakness in my right arm and leg), which come and go, for the past three years. A neurologist suspected multiple sclerosis, but all the tests were negative and no other cause has been found.

Diana, Illinois

Hi! I have just finished reading the article on "Endometriosis and Yeast" and had to write to you immediately. I found this article very encouraging. Let's hope this research is on the right track.

As I went through the list of possible connections between endometriosis and candida, I was amazed at how many symptoms I could relate to. I am very allergic to many things—trees, grass, weeds, etc., but also all antibiotics. I was given tetracycline when I was 16 for the treatment of acne and at 17 I had a very bad case of mononucleosis and was treated with cortisone.

When I was 19, I went on the birth control pill. The first few years were fine, but I started getting terrible cramps and was in pain for two weeks every month. I went to many doctors who all said it just couldn't be so because the pill was supposed to take away cramps, not give them. They all said it was in my head . . . Soon after I went on the pill, I developed a spastic colon. I suffered for 11 years with terrible diarrhea, and then spells of constipation . . .

I also went to many doctors regarding this problem, and they said it was my nerves and put me on various medications. Through my own research, I discovered I was allergic to eggs and milk. My diarrhea is much worse when I eat anything with eggs or milk in it.

I also can relate to the point about urgent or frequent urination. I just thought I had the weakest bladder in the world. My husband is amazed at how often I have to urinate. I must get up 3–4 times during the night.

I have a terrible problem with nasal congestion, which I contributed to my allergies. This problem is getting worse, and is going into my chest and feels like asthma because I can't get any air in and I wheeze . . . I feel worse on damp days and get very congested. I constantly crave sweets and alcohol (but I seem to not be able to tolerate it well, and I don't feel very well even after only a few glasses of wine), and I can't stand being around people who smoke because it does make my symptoms worse.

I'm infertile due to my endometriosis and spend my entire time going to specialists, so of course, I get depressed . . . I think the worst part of all this is how the medical profession has treated me. I have been to more doctors than I like to remember and they never seem to take my situation seriously. It is very frustrating.

MY COMMENTS

Countless women with endometriosis have found relief and returned to health with the help of Mary Lou Ballweg and the Endometriosis Association. The yeast connection to this debilitating disease is becoming better established all the time. I recommend that women with endometriosis that is not responding to conventional treatment discuss the possibility of candida infection with their physicians. While candida is probably not responsible for all endometriosis, it plays a large role in many cases.

RESOURCES

- My website: www.yeastconnection.com has links to the best natural supplements to help you deal with endometriosis and other yeast-related problems.
- Endometriosis Association website: www.endometriosisassn.org, (800) 992-3636 (for a free packet of information). For other information, call (414) 355-2200

REFERENCE

1. Rier S. E. et al. Endometriosis in Rhesus monkeys (Macaca mulatta) following chronic exposure to 2,3,78,-tetrachlorodibenzo-p-dioxin. Fundamental and Applied Toxicology. 1993(21);433-441.

CHAPTER 14

Vulvodynia

You may never have heard the term vulvodynia, but many of you will recognize the symptoms: burning pain, stabbing pain, itching, stinging and irritation around the vulva, often accompanied by a white cheesy discharge.

One patient described her external genitalia as "bright red," and another said it "feels like it's on fire."

The pain can come suddenly, seemingly from nowhere. It may be constant or intermittent. It may take place only during sexual intercourse. For some women, it is more pronounced near their time of their menstrual periods.

Frequently, the area around the vagina shows no redness or swelling or infection, in spite of these intense symptoms.

The pain can involve the entire area, including the rectal or anal skin and may include hypersensitivity around the small labia, making it difficult to walk. Other women may experience discomfort when their pubic hair is touched, making it difficult for them to wear underwear.

Occasionally women also experience a painful or hypersensitive clitoris.

The medical community has little awareness of this potentially disabling condition. It is common for patients to go from doctor to doctor for years before they get help. It's equally common for doctors to suggest there is a psychological cause for the condition.

I had never heard of vulvodynia until early 1993, when a Nashville psychiatrist asked me to see one of his patients, a professional woman who had been troubled by persistent intense burning of her external genitalia for several years.

It was this patient, whom I'll call "Laura" (not her real name), who, through her own research on her condition, opened the door for me to a whole new element of dysbiosis.

In fact, the term "vulvodynia," meaning literally "burning vulva," was coined in the mid-1980s in response to what physicians began to see as a syndrome affecting many patients.

A pioneer in recognizing and treating vulvodynia is Dr. Marilynne McKay, M.D., who was, before her retirement, a professor of dermatology at Emory University in Atlanta. Dr. McKay recognized the level of frustration many women who suffered from vulvodynia were feeling. In a comprehensive review article published in 1989, she wrote:

> Because physical signs may be subtle, many have been told that their problem is primarily psychological, especially when dyspareunia (difficult or painful sexual intercourse) is a major component. Unrealistic expectations and unrewarding medical experiences contribute to the resentment, frustration and anger so often expressed by these patients.[1]

The National Vulvodynia Association released some staggering information from a population-based study in late 2001. It showed:

1. Nearly 20% of all American women have suffered from chronic vulvovaginal pain at some point in their lives.
2. But 40% of these women never sought treatment for the condition, many of them saying it was too "embarrassing."
3. And 40% of those who did seek treatment never received a definitive diagnosis.

From the many patients I know with this problem, I am sure these feelings will be uncomfortably familiar.

Soon after learning about the term vulvodynia, I received many calls and letters from women with this syndrome. Then as I thought about it, I realized that many of the women with yeast-related problems who said, "I'm troubled by a constant yeast infection," were, in fact, suffering from vulvodynia.

In a later article, Dr. McKay explains that bacterial, fungal and viral infections should all be considered as causes for vulvodynia.

Candida is by far the most important infectious agent to consider in the evaluation of patients with vulvodynia.[2]

WHAT CAUSES VULVODYNIA?

There are a number of theories about the cause of vulvodynia. Among those listed by the National Vulvodynia Association:

- an injury to, or irritation of, the nerves surrounding the vulva
- an abnormal response of different cells in the vulva to environmental factors (such as infection or trauma)
- genetic factors associated with susceptibility to chronic vestibular inflammation
- a localized hypersensitivity to candida (yeast)
- spasms of the muscles that support the pelvic organs

Others, myself included, believe many cases, perhaps one-third or more, of vulvodynia can be one symptom of a systemic yeast overgrowth.

TREATMENT

I have some information from the Vulvar Pain Foundation that suggests many vulvodynia patients have excess oxalate in their urine. This is the same material that produces kidney stones in its extreme form and sometimes causes crystals in the urine. A diet low in oxalates and supplementation with calcium citrate and equal amounts of magnesium citrate has produced good results for some people. More information on the low oxalate diet (that is, in many ways, similar to the anti-yeast diet) is available from the VPF at its website:

www.vulvarpainfoundation.org

or by writing:
VPF
P.O. Box 177
Graham, NC 27253
(a self-addressed, stamped envelope is appreciated)

The VPF's literature also contends that many women who have vulvodynia have multiple health problems, which sounds very much like systemic candida overgrowth.

Some physicians prefer to treat the yeast infection that frequently accompanies complaints of vulva pain, and then investigate further if it continues after the yeast problem has been successfully addressed.

Dr. Paul Nyirjesy, M.D., a member of the board of the National Vulvodynia Association and an associate professor of obstetrics and gynecology at Jefferson Medical College in Philadelphia, subscribes to the "eliminate the yeast" plan. In a conversation with him he said,

> In my experience, about 50% of women with vulvodynia vestibular syndrome (VVS) will at some point during their care with us have a positive culture for yeast. This 50% is greater than the 15–20% culture positivity rate in the general population.
>
> Many women with this syndrome respond to antifungal medications, but there is a small group of women who will suffer recurrent yeast infections, defined as four or more infections in a 12-month period.
>
> Current studies suggest that in many women with recurrent yeast infections, local immune factors play a key role, and that patients get into a yeast growth spiral they can't break without immune system support.

This same doctor admits that many of his colleagues don't agree with him and his philosophy of "culture everyone for yeast and culture often." He also said:

> I am constantly struck by the number of patients that we see who have initial negative cultures, then get a positive one, and finally see all of their symptoms improve once they get started on antifungal therapy. Many patients have found positive results with 100–200 mg. of Diflucan once or twice a week for six months.

Dr. Nyirjesy says he uses Diflucan almost exclusively if there is a positive culture for candida.

"I've treated well over 1,000 patients with long-term Diflucan. It's well tolerated, and maybe once every two years I have a patient who gets a positive culture while she's being treated. But for the vast majority, they have negative cultures while they're on treatment," says Dr. Nyirjesy.

MY COMMENTS

As most women with yeast-related problems have found out, there's no magic cure—no "quick fix" for many of their recurring health problems. These include fatigue, muscle aches, headaches, vaginitis—and vulvodynia. Although I've focused on the "yeast connection" to all of these symptoms, I've also said there are multiple causes, and antifungal medications won't provide answers for every woman. These conditions are frustrating not only for the people with these disorders, but also for the health professionals who care for them.

I've emphasized in this book that every part of a woman's body is related to every other part. I was pleased to read the comments of the Vulvar Pain Foundations' Scientific Research Committee that said that interstitial cystitis, fibromyalgia, vulvodynia and irritable bowel disorder are directly related to each other.

REFERENCES

1. McKay, M., "Vulvodynia—A Multifactorial Clinical Problem," *Arch. Of Dermatology,* 1989; 125:256–262.

2. McKay, M., "Vulvodynia Diagnostic Patterns," *Dermatol. Clin.,* 1992, 10:423–433.

PART 3

Yeast-Related Problems That Affect More Women Than Men

CHAPTER 15

Headaches

More than 45 million Americans have chronic, recurring headaches, and 62% of these are chronic migraine sufferers, according to the National Headache Foundation.

Both sexes are affected, but women bear the brunt of headaches, particularly migraines, suffering from them three times as often as men.

There are two major types of headaches:

- **Tension-type:** These include 90% of all headaches, characterized by tightening of the muscles in the neck and back of the head, where the pain usually occurs.
- **Vascular headaches:** These are believed to be caused by an interaction between blood vessels and nerves in the brain covering. This category includes migraines and cluster headaches.

The problem is so widespread and debilitating that it costs American industry $50 billion a year due to worker absenteeism.

Interestingly, there is some evidence that suggests that women who take birth control pills are more likely to suffer chronic headaches and migraines or to have their headaches worsen.

The National Headache Foundation says most migraine sufferers are between the ages of 20 and 45, and the condition is often hereditary: "In fact, a child has a 50% chance of becoming a sufferer if one parent suffers, and a 75% chance if both parents suffer."

SYMPTOMS

Migraines are characterized by throbbing head pain, usually located on one side of the head, often accompanied by nausea and light and sound sensitivity.

There are many theories about the causes of migraines, but many sufferers say attacks can be triggered by certain types of food (including aged cheeses, chocolate and red wine), changes in climate, emotions, medications and hormones. The hormonal component may help explain why women are more prone to headaches and migraines, since they are subject to more hormonal swings than men.

TREATMENT FOR MIGRAINES

There are medications available for the treatment of migraines, and some of those medications help with certain types of migraines and others do not.

There are preventive medications, taken every day, that can help reduce the number of attacks, and abortive therapy that helps stop the symptoms of a migraine after the attack begins.

The connection between headaches and allergies goes back decades. In fact, the earliest connection was made 84 years ago.

In my pediatric and allergy practice, many of my patients (including children and adults) found relief from their headaches when they avoided a common food in their diets. I published my observations for the first time in a report on 50 patients in 1961. The title of the article was "Systemic Manifestations Due to Allergy." Although fatigue, mental and emotional symptoms were present in 49 of the 50 patients, some 26 complained of headaches. In 10 patients, headaches were a major complaint, and in 16, they were a minor complaint.[1]

In a subsequent article in 1975 in the *Pediatric Clinics of North America,* I said:

> I see a lot of youngsters who complain of headaches. Not infrequently, some of them have been to family physicians, ophthalmologists, otorhinolaryngologists (ear nose and throat specialists) and neurologists. And when such children have been found to have nor-

mal blood pressure, normal sinus x-rays, normal skull x-rays and elec-
troencephalogram, normal vision and normal findings on physical
and laboratory examinations, either the mother or physician is apt to
conclude, "Johnny's complaining of headaches to get attention. He
must be suffering from an emotional problem."[2]

In this article, I cited the observations of several leading allergists,
including the late Jerome Glaser, who published an article in 1954 entitled, "Migraine in Pediatric Practice," and John T. McGovern and T. J.
Haywood, who wrote a chapter on allergic headaches in Frederic
Speer's classic book, *Allergy of the Nervous System.*

Several years later in a commentary in *The Lancet* entitled "Food
Allergy," a British physician, Dr. Ronald Finn, said:

> The concept that certain foods can produce abnormal reactions in
> susceptible individuals has a very long history.

Dr. Finn also reviewed reports in the medical literature dating from
1919 by researchers who noted that food allergens can be absorbed
through the digestive tract. He also referred to the pioneer observations
of Albert Rowe and Theron Randolph. Dr. Finn explained that immediate swelling and itching are easily recognized symptoms of an allergic
reaction, and the patient can then easily avoid the offending food. But,
he noted:

> Food allergy reactions are usually more subtle, and the patient does
> not realize the cause of these symptoms. Indeed, food allergens are
> usually favorite foods which are eaten regularly, often in excess.[3]

 See www.yeastconnection.com for a printable weekly form to help
you identify foods you eat regularly or in excess.

In 1979, another British physician, Dr. Ellen Grant, published the
findings of her study of 60 migraine patients. The majority of her patients were women whose average age was 39. The mean length of migraine history was 18 years.

Most patients showed other symptoms, including lethargy, depression, anxiety, dizziness, abdominal pain and menstrual problems. Others gave a history of recurrent infections, especially cystitis.

In studying her patients, Dr. Grant noted that a migraine was sometimes precipitated by foods containing amine, especially cheese, chocolate, citrus fruits and alcohol. When her patients discontinued these foods, there was a highly significant decrease in the number of headaches. Yet, only 13% were completely headache-free.

In studying 186 migraine patients, she advised them to use elimination/challenge diets similar to the diet I recommend. Some 126 attempted it and 85% became headache-free. The most common foods causing reactions were wheat (78%), oranges (65%), eggs (45%), tea and coffee (45% each), chocolate milk (37%), beef (35%), corn, cane sugar and yeast (33% each).[4]

Another British researcher, Dr. Joseph Egger, placed his headache patients on a limited food diet and found that 93% of 88 patients experienced relief.[5]

Drs. Jean Monro, Jonathan Brostoff and associates studied a group of British patients with severe migraines, which they said affect about 20% of the population, with more women suffering than men. They found that two-thirds of the people with migraines/severe headaches were allergic to certain foods as shown by an elimination diet and subsequent challenge.[6] (In an elimination diet, certain foods are eliminated for a period of five to 10 days. Patients observe changes in their symptoms during that time. If they improve during the five-to-10-day elimination, they then challenge, or reintroduce, one food at a time to see if symptoms reappear.)

U.S. investigators who have noted the diet/migraine connection include O'Banion, Bahna and Mansfield. In a 1986 article, Mansfield said:

A large body of literature supports a role for food allergy as a cause of migraine.

He cited reports in the medical literature as early as 1977. He also reviewed clinical reports by Vaughn, published in *The Journal of the American Medical Association* in 1927 and reports of others published in subsequent years. I was especially struck by this paragraph from Mansfield's report:

> In 1952, Unger and Unger published a manuscript in the *Journal of Allergy* entitled, "Migraine is an Allergic Disease." Thirty-two years later, Monro, Carini and Brostoff would publish an article entitled, "Migraine is a Food Allergic Disease." Such is the pace of progress.[7,8,9,10]

HEADACHES AND THE YEAST CONNECTION

Several years ago when I analyzed the histories of 100 consecutive women with yeast-related problems, fatigue, headaches and depression were the most common complaints. And during the past 10 years, I've received countless letters from people whose primary complaint is headaches.

I've also met and talked to many people who have described yeast-related headaches. On a trip to California in 1992, I met a television personality (I'll call her Reba) who told me that her headaches were yeast-related. And I was naturally pleased when she said that my book, *The Yeast Connection,* had played a major role in helping her regain her health and her life. So I asked her to write me a letter summarizing her story.

Reba's Story

I moved to Los Angeles in 1979 and started developing headaches, even though I'd never been a headache sufferer before. I went to every allergist at UCLA. No one could find the source of my allergic rhinitis, which they said was my problem.

Two years later I moved to New York and I continued to have debilitating headaches. I took all sorts of over-the-counter medications. Nothing helped. So I kept on seeing specialists who were convinced I was crazy and were prescribing mood elevators—but I wouldn't take them.

Finally I happened upon a nutritionist who said that I was a classic candidate for the yeast connection and told me to go out and buy your book. He told me to stop eating bread and dairy immediately. He also put me on vitamins, minerals and acidophilus.

I improved some, but he said, "As long as you keep drinking coffee and insist on eating sugar, you'll never get rid of your headaches." So I followed his instructions and today, five years later, I can tolerate wheat and bread in limited quantities and take occasional sugar. I continue to stay away from dairy.

I never did have to take prescription yeast drugs. I did it mainly with a diet and acidophilus and lifestyle changes.

I also had a secretary working for me last year who had headaches all the time. She started describing her symptoms to me. Her father thought she was a hypochondriac and told her to stop being hysterical and told her she needed to see a shrink.

Because her symptoms were so much like mine, I felt she had yeast problems. I'm happy to report that she found a naturopath here in L.A. and he started treating her for yeast. She's fine now.

MY COMMENTS

How and why are headaches and yeast connected? There are probably several mechanisms, as described elsewhere in this book. But it seems to me that a principal one is the disturbance in the balance of normal bacteria in the intestinal tract, which leads to a "leaky gut," and the absorption of food antigens and toxins throughout the system. Those toxins may be connected to allergies and food sensitivities, particularly to foods that are migraine triggers.

REFERENCES

1. Posner, J. D., M.D., Cecil's Textbook on Medicine, W. B. Saunders Co., Philadelphia, PA, 1988; p. 2129.

2. Crook, W. G., Harrison, W. W., Crawford, S. E. and Emerson, B. S., "Systemic Manifestations Due to Allergy," *Pediatrics,* 1961; 27:790–799.

3. Finn, R., "Food Allergy," *The Lancet,* February 3, 1979.

4. Grant, E. C., "Food Allergies and Migraine," *The Lancet,* 1979; 1:986–988.

5. Egger, et al, "Is Migraine Food Allergy? A double-blind control trial of oligoantigenic diet treatment," *The Lancet,* 1983; 2:865–868.

6. Monro, J., Carini, C., Brostoff, J., "Migraine Is a Food Allergic Disease," *The Lancet,* 1984; 2:719–721.

7. O'Banion, D. R., "Dietary Control of Headache Pain, Five Case Studies," *Journal of Holistic Medicine,* 1981;3(2)2, 140–150.

8. Bahna, S., "Food and Additive Sensitivity Present Diagnostic Dilemma," *Consult—A Forum for Physicians,* The Cleveland Clinic Foundation, 1988; Vol. 7 No. 3, pp. 8–9.

9. Mansfield, L. E., "The Role of Food Allergy in Migraine: A review," *Immunology and Allergy Practice,* 1986; Vol. 8 No. 12, pp. 406–411.

10. Vaughn, W. T., "Allergic Migraine," *JAMA,* 1927; 88:1383–86.

CHAPTER 16

Depression

Approximately 19 million Americans suffer from clinical depression, and two-thirds of them are women.

Nearly 15 years ago, when I decided to review the records of 100 of my patients with yeast-related conditions, I was surprised at the results: depression, fatigue and headache were the most common complaints. What's more, 85% of these patients were women and most of them were between the ages of 30 and 45.

I had listened to these patients. I knew that the multiple health problems typical of people with candidiasis could become overwhelming, but those numbers gave me enough of a jolt to deepen my search for the link between depression and *Candida albicans.*

In his lectures and medical reports, Dr. Truss discussed the many and varied manifestations of patients he had seen with candida-related health problems. And in his book, *The Missing Diagnosis,* he discussed many complex health problems that he had found to be yeast-related.[1] Included were hormone dysfunction, depression and manic depression (now called bipolar disorder). And he said:

> The mechanism of depression is poorly understood. Certain chemicals are known to be involved in normal brain function. It is thought that depression occurs when some factor upsets their proper balance or interferes with their proper function.

Dr. Truss also pointed out that some patients with depression respond to anticandida therapy—often dramatically. But he cautioned:

Depression is a serious and potentially dangerous condition and one deserving care by a competent psychiatrist; self-diagnosis and treatment should never be attempted. It is perfectly reasonable to look for some correctable cause of depression, but even when found, its treatment should not immediately and abruptly replace the psychiatric program.

Drugs prescribed by the psychiatrist should be gradually withdrawn, preferably under his supervision. Their use should not be discontinued suddenly or prematurely.

SYMPTOMS

The National Mental Health Association lists the following symptoms of depression:

- Persistent sad, anxious, or "empty" mood
- Sleeping too little, early morning awakening or sleeping too much
- Reduced appetite and/or weight loss, or increased appetite and weight gain
- Loss of interest in activities once enjoyed, including sex
- Restlessness, irritability
- Persistent physical symptoms that don't respond to treatment (such as headaches, chronic pain or digestive disorders)
- Difficulty concentrating, remembering, or making decisions
- Fatigue or loss of energy
- Feeling guilty, hopeless or worthless
- Thoughts of suicide or death.

Depression and bipolar disorder can develop from many different causes. These include genetic factors, nutritional deficiencies, endocrine disturbances and psychological stress or trauma.

MEDICAL SUPPORT FOR THE RELATIONSHIP OF CANDIDA TO DEPRESSION

Several years ago, John W. Crayton, M.D., professor of psychiatry at Loyola Medical School in Chicago, described his laboratory findings in a group of patients with fatigue, weakness, depression and many other symptoms. The 28 subjects who were studied ranged in age from 18 to 45; 20 were women and eight were men. All gave a convincing history of an increased intensity of symptoms after they ate.[2]

Antibody studies in these patients showed higher levels of candida antibodies than in a control group without symptoms.

In studies carried out at the University of Tennessee several years ago, Jay Schinfeld, M.D., studied a group of women with severe premenstrual syndrome (PMS) and a history of vaginal candidiasis. And he noted that "depression was often found in women with severe PMS."[3]

Various treatment programs were tried in managing these patients, including oral nystatin and yeast-elimination diets. Although Dr. Schinfeld's study was a small one, patients who received anticandida therapy showed significant physical and psychological improvement when compared to a group who didn't receive the therapy.

In May 1996, two Boston psychiatrists reported case histories of two patients with chronic depression who were effectively treated with ketoconazole (Nizoral).

The first of these patients was a 44-year-old woman with a seven-year history of chronic depression, and the second patient was a 35-year-old woman with a history of chronic atypical depression since childhood. In each of these patients, the depression and associated symptoms, including decreased libido and fatigue, responded favorably to the administration of ketoconazole (Nizoral).

In their concluding paragraph, these observers said ketoconazole appeared to have some value for patients with chronic depression, especially for patients who cannot tolerate the side effects of commonly used antidepressants, and called for further study.[4]

In her book, *Solving the Depression Puzzle,* Rita Elkins, M.H., covers triggers and causes of depression, including candida. She lists many therapies that help fight depression, including low-sugar diets, nutritional supplements like 5-HTP, magnesium, Omega-3 fatty acids, St. John's wort, SAM-e plus exercise and acupuncture.

Clinical research has shown SAM-e (s-adenosyl methionine), a natural substance present in all cells, to be effective not only against depression, but also in treating the muscle and joint pain of fibromyalgia.

COMMENTS FROM DR. CAROLYN DEAN

In the years since Dr. Crook's death, there have been new discoveries as well as interesting new theories about the connection between yeast and depression. Of the more than 180 toxic byproducts of candida yeast that have been identified, two, alcohol and acetaldehyde, are known to be particularly harmful. Measurable levels of alcohol and acetaldehyde in the blood have been found in some patients with severe yeast overgrowth.

When acetaldehyde reacts with the brain-chemical dopamine, it can cause feelings of anxiety, depression, poor concentration and a sense of being "spaced-out."

Thus it should be no surprise that depression and fatigue are the most common complaints of people suffering from yeast overgrowth.

Another of candida's predominant toxins, *canditoxin,* depletes essential dietary nutrients and has also been associated with behavioral changes.

Clearly, without proper treatment for candida, a multitude of serious problems arise that can trigger even deeper depression.

Here is why this is a vicious cycle:

- Candida toxins directly affect your brain, leading to feelings of depression.
- Candida overgrowth causes sugar and carb cravings, which can cause your diet to spin out of control.
- More nutritional deficiencies develop.
- More endocrine disturbances occur.
- Your immune system is further weakened.

This promotes the release of certain brain chemicals that disturb the brain's normal mood-regulating balance, triggering increased anxiety or depression.

Remarkably, despite overwhelming anecdotal evidence from sufferers, there has not been any significant research on yeast-related causes of depression over the past 20 years—since Dr. Crook introduced the topic in an article he wrote for the *Journal of the American Medical Association* (JAMA).

While there is no doubt in my mind that yeast overgrowth can and does cause imbalances in brain chemistry that will trigger depression, the happy news is that our 6-Point Yeast Fighting Program can ease depression, even if you've suffered for years. If you've battled with depression, you'll be interested in Michelle's story, told in her own words in Chapter 36.

COMMENTS FROM DR. CROOK

Over the span of my medical and writing career, I have received thousands of moving letters from women suffering from depression. Only within the past several years have I begun to receive reports from women with yeast-related bipolar disorder, in which the affected individual experiences powerful mood swings that range from "very elevated" mood (mania) to deep depression.

I think the story below, told by a New Zealand woman I call Martha, is especially heartrending:

Martha's Story

Until I was 38, I never knew what depression was apart from a couple of very brief bouts. I remember once coming out with those terrible words, "What on earth has she got to be depressed about?"

Then it happened. I developed a pattern of manic depression, swings of mood from extreme highs to extreme lows. It lasted nine years and, ultimately, nearly cost me my life, smashed up my career and could easily have finished my marriage. It caused me unspeakable suffering.

Depression was like I imagined it would be having both legs amputated or losing an adored child—I often thought if it were to go on like this I'd rather be dead. But in the end, the prisoner without hope was rescued from the dungeon, so this tale has a happy ending.

Martha's illness started with a viral infection that caused liver damage. Although she apparently recovered from this illness, she developed recurrent episodes of mood swings. Two and a half weeks up, two and a half weeks down—that completely took over her life. Describing how she felt when depressed, she said:

Imagine the whole world spray painted gray or being in a small windowless cell or in a tunnel. I had no energy or drive whatsoever. I used to feel that I had 50-pound weights on each foot and—30-pound weights on each wrist. All my favorite things suddenly became meaningless and sterile.

If someone had given me two round-trip air tickets to London and Paris and $15,000 spending money, I would have been completely unmoved. Nothing could trigger a flicker of interest or enthusiasm.

. . . If Yul Brenner himself had purposely moved over to my side of the bed, I would have rebuffed him! I was sexually 100% dead.

Then, after a week or two of this hell, Martha would swing into a manic world of exhaustion and even delirium.

I was king of the castle, drunk with joy, bursting with crazy schemes, on the go literally 22 hours a day, talking nonstop, constantly interrupting, spending money like water, issuing dinner and party invitations, smashing up the car. You'd have to have seen it to believe it . . .

I was diagnosed from the beginning as a "textbook case of manic depression." Over the whole depression period, I had been sent to a string of different specialists for what seemed quite unrelated conditions.

After struggling with this problem for eight years, Martha began to deteriorate rapidly and was put in a psychiatric hospital. Following discharge, she received therapy outside of the hospital, including medication and group therapy for many weeks and months.

Martha continued:

In spite of all the tears and the new insights, the manic depression didn't go away . . . Then came the great breakthrough I'd waited so many years for. My general practitioner prescribed eight nystatin tablets a day for the candida, plus large doses of the B vitamins and calcium. The bouts of mania stopped immediately, the depressions became briefer and much less severe; within three months, they disappeared.

In addition, Martha found that a number of foods, including sugar, white flour and eggs, played a part in causing her symptoms. In her concluding comments she said:

I can hardly believe it's really over. I feel tremendous gratitude that I've been saved from this living death. But I also feel angry that I wasn't properly diagnosed earlier . . . Of course, I can only assume that the Candida albicans *and its treatment with nystatin was the critical thing in my case. I can hear the doctors say, "You would have recovered anyway." But to me, it is beyond a reasonable doubt that candida/nystatin was the answer. It was the only new factor after nine years of illness.*

Now that I have been completely free of manic depression for a year, I plan to visit all the psychiatrists and specialists involved in my treatment over the past nine years to tell them what happened to me. Please, God, my sufferings might help others even in a small way . . . the medical profession should get this sort of feedback.

MY COMMENTS

Depression and manic depression, like chronic fatigue syndrome, fibromyalgia and other disorders described in this book, can develop from many different causes. These include genetic factors, nutritional deficiencies, endocrine disturbances, viral infections, chemical sensitivities and toxicities and psychological stress or trauma.

I do not want you (or anyone) to think I'm saying that *Candida albicans* is the cause of depression. Yet, if you suffer from depression and/or any other disabling disorder and have a history of the following:

- repeated or prolonged courses of antibiotic drugs
- persistent digestive symptoms
- and/or recurrent vaginal yeast infections

a comprehensive treatment program that features oral antifungal medications and a special diet may enable you to change your life.

For my patients, antifungal medication and a sugar-free special diet did not necessarily provide a "quick fix," but 85% of them improved significantly following the anticandida program.

RESOURCES

For more success stories, see my book, *Yeast Connection Success Stories,* Professional Books 2002.

Cass, Hyla, *Natural Highs,* Avery, 2002.
Elkins, Rita, *Solving the Depression Puzzle,* Woodland Publishing, 2001.

REFERENCES

1. Truss, C. O., *The Missing Diagnosis,* P.O. Box 26508, Birmingham, AL 35226, 1986; pp. 73, 75–76.

2. Crayton, J. W., "Anticandida Antibody Levels in Polysymptomatic Patients," Candida Update Conference, International Health Foundation, Memphis, Tennessee, September 16–18, 1988.

3. Schinfeld, J., "PMS and Candidiasis: Study explores the possible link," *The Female Patient,* 1987;(12):66–70.

CHAPTER 17

Chronic Fatigue Syndrome, Fibromyalgia and Hypothyroidism

I f you're tired, so tired and add the "sick all over" complaint to the laundry list of your health problems, it's likely you've been from doctor to doctor with no real help and probably one or more of those doctors has suggested, "It's all in your head."

It may be in your head, but it's probably also in your reproductive tract, your digestive tract, your joints and muscles, and just about everywhere in your body.

Chronic fatigue syndrome, fibromyalgia and hypothyroidism, disorders that have received a great deal of attention in recent years, are each typified by a huge list of symptoms for which there are few real remedies.

And while they are three different diseases, they are often spoken of in the same breath and may be connected. I'm lumping them together in this chapter because they are so similar; however, I'll talk about each of them separately for clarity.

I suggest to you that some cases of chronic fatigue and immune dysfunction syndrome (CFIDS) and fibromyalgia syndrome (FMS) may be connected to yeast overgrowth for the simple reason that a significant number of patients find relief when they follow the anti-yeast diet, probiotics and antifungal medication regimen.

I developed an interest in food-related chronic fatigue, headaches, muscle aches, irritability and other symptoms when I was still in pediatric practice more than 30 years ago.

At that time, an alert mother convinced me (against my will) that her 12-year-old son, Tom's, fatigue, weakness, inability to get out of bed in the morning and other symptoms vanished when he stopped drinking milk.

A short time later, I read several articles in the medical literature that described food-related fatigue and other symptoms. After reading these articles, I put a number of my patients on five- to 10-day elimination diets, followed by challenge (re-introducing certain foods one at a time to the diet to determine if they sparked the symptoms anew).

Although I didn't help *all* of these tired, irritable, unhappy patients, I was excited when I received reports from mothers who said, "Susie's like a different child. But when I give her chocolate milk, corn chips, wheat or eggs, her symptoms return."

During the 1970s, I began to see more adult patients, especially women, who complained of fatigue and other symptoms that made them feel "sick all over." Some of these patients were helped by focusing on allergy treatment, changing their diets and cleaning up the pollutants in their homes. Yet, many remained ill.

Then in the late-1970s and on through the 1980s, I found that many of my patients with chronic fatigue and other symptoms improved, often dramatically, when they changed their diets and took nystatin or Nizoral.

During the mid-1980s, a small but growing number of physicians found that a special diet and nystatin, Nizoral or Diflucan, helped many of their patients, including some with chronic fatigue syndrome.

CHRONIC FATIGUE SYNDROME—AN ENIGMA

Sometimes this mishmash of symptoms is called CFS (chronic fatigue syndrome), while others call it CFIDS (chronic fatigue and immune dysfunction). The two terms are interchangeable.

The Chronic Fatigue and Immune Dysfunction Syndrome Association of America says 800,000 Americans have the disease, but no more than 16% have been diagnosed. "The myth that CFIDS is the 'yuppie flu' and is most prevalent in young, well-to-do professional, white females has not been supported by research. Recent community studies show that, in reality, persons of color and lower income are at greater risk for CFS."

The U.S. Centers for Disease Control and Prevention (CDC) defines CFIDS as "unexplained fatigue of greater than or equal to six months' duration."

Symptoms

For an affirmative diagnosis of CFIDS, the CDC says four or more of the following symptoms must occur at the same time:

1. debilitating fatigue
2. post-exertional malaise lasting more than 24 hours
3. impairment in short-term memory or concentration
4. sore throat
5. tender lymph nodes
6. muscle pain
7. multi-joint pain without swelling or redness
8. headaches of a new type, pattern or severity
9. unrefreshing sleep

The CFIDS Association adds, "CFIDS brings with it a constellation of debilitating symptoms . . . It is characterized by incapacitating fatigue experienced as profound exhaustion and extremely poor stamina."

The group points out that CFIDS is not a psychological disorder.

Treatment

There are no diagnostic tests for CFIDS, and no cure has been found, although there are a number of treatments outside the yeast-connected treatments that may or may not help, including:

- medications for the digestive problems and nausea that often accompany CFIDS
- medications for depression and anxiety
- counseling to help develop coping skills necessary to live with a debilitating chronic disease
- gentle exercise
- changes in diet
- sleep and rest management

FIBROMYALGIA SYNDROME (FMS)

In an article, "Fibromyalgia—An Emerging But Controversial Condition," Don L. Goldenberg, M.D., of the Tufts University School of Medicine in Boston, described the symptoms, laboratory findings and treatment results of 118 patients with fibromyalgia. Here are excerpts from his article:

> Fibromyalgia is one of the most common diagnoses in ambulatory practice. Recent estimates of the instance of fibromyalgia in the United States have ranged from 3 to 6 million . . . Common symptoms noted in our series and other recent reports have included neck and shoulder pain, morning stiffness, sleep disturbances and fatigue.
>
> Seventy to 90% of patients have been women, and the mean age for diagnosis has varied from 34 to 55 years. The average duration of symptoms at the time of diagnosis has been five years. Most patients reported little change in their symptoms (while taking various medications) . . . and continue to have symptoms whatever medication is used.

In a later article published in 2002, Dr. Goldenberg affirmed that some of the confusion around the difference between fibromyalgia and CFIDS may not be terribly important:

> For example, fibromyalgia, chronic fatigue syndrome, irritable bowel syndrome, chronic headaches and low back pain are all, I believe, interrelated conditions that may share a number of overlapping symptoms. I try to make people understand that I'm not as concerned about differentiating fibromyalgia from, say, chronic fatigue syndrome,

as I am about excluding more defined and potentially dangerous conditions, such as rheumatoid arthritis, systemic lupus, hypothyroidism and others.[1]

The symptoms of fibromyalgia are elusive because they can change between one visit to the doctor and the next. That elusiveness probably contributes to the difficulty in reaching a diagnosis.

Pain is by far the most prominent symptom of fibromyalgia. Some people describe it as "knife-like" or a "muscle cramp." Most people say the pain never goes away, and it can be aggravated by stress, anxiety, physical overexertion, weather changes, hormonal changes and depression. Of course, that pain disrupts sleep patterns for 75% of the sufferers and results in extreme fatigue for 90% of the people with fibromyalgia.

Symptoms

Among the other symptoms (in alphabetical order) are:

- abdominal pain
- bladder irritability or spasms
- blurred vision
- chest pains and pressure beneath the breast bone
- cramps
- dry eyes and mouth
- falling
- fatigue
- gastroesophageal reflux (heartburn or acid stomach)
- general aches and pains
- hearing loss
- intermittent hearing problems and low-frequency hearing loss
- memory and reasoning problems (brain fog)
- migraine or tension headaches
- morning stiffness
- muscle twitching
- nighttime grinding of teeth (bruxism)
- pelvic pain
- pre-menstrual syndrome

- skin sensitivity to temperature
- sleep problems, insomnia
- temporomandibular joint disorder (TMJ)
- tingling or numbness in arms, legs, feet or face
- water retention or swelling, especially in the hands, face and feet

Cause

At several conferences I attended on chronic fatigue syndrome, professionals discussing this disorder expressed varying points of view, and different treatments were recommended. The general consensus seemed to be that FMS and CFIDS were closely related—if not the same disorder.

I met Kristen Thorsen, president of the American Fibromyalgia Association, at a conference in California about 10 years ago. Since then, we have corresponded and talked on the phone a number of times. She says, "FMS could be another name for CFIDS . . . or vice versa . . . depending on which diagnosis you happen to have." Because of her own problems with FMS, for a number of years she's published *Fibromyalgia Network,* a newsletter for people with FMS/CFIDS.

Here are excerpts from the newsletter:

- It's hard to state the cause of FMS/CFIDS. Genetic predisposition combined with triggers can lead to the development of the disease. These triggers also continue to work as aggravants once the condition reaches full-blown development.
- Trauma (especially to the neck or spinal cord region) and candida and bacterial infections may be triggers.
- Infectious agents have the ability to generate the production of nasty pro-inflammatory cytokines.
- What is FMS? A CNS (central nervous system) disease that also causes dysfunction in the ANS (autonomic nervous system) which in turn feeds back to the CNS problem.
- A dysfunction in CNS and ANS can lead to thyroid and adrenal problems and other hormonal abnormalities.
- Sleep disturbances are common. It is unknown whether sleep disturbances such as Periodic Limb Movement Syndrome (PLMS),

also known as restless legs syndrome, during sleep are secondary or primary to the development of FMS.

- Immune system. When sleep is destroyed, the immune system is adversely affected, and infections become a problem. Yeast overgrowth can have devastating effects. So can other infections, but antibiotics given for them promote the overgrowth of yeast, which is known to cause cytokine production and pain.
- Small intestinal bacterial overgrowth (SIBO) can be a trigger and cause problems.

As in the case of CFIDS, there is no laboratory test to diagnose fibromyalgia.

Treatment

Standard treatment options for fibromyalgia include:

- Antidepressant medication, particularly tricyclic antidepressants, such as Elavil
- Another class of antidepressants called serotonin-reuptake inhibitors (SSRIs), like Prozac, which may help lift mood, but don't address the physical condition. They are also associated with a growing list of side effects
- Pain relievers
- Trigger point injections that involve injecting a local anesthetic such as lidocaine in the tender point
- Lifestyle changes, including exercise, establishing sleep routines and relaxation techniques (meditation, biofeedback and yoga), and avoiding caffeine

In addition, the American Fibromyalgia Association recommends:

- antibiotics/antifungals for infections
- trigger-point injections combined with massage therapy and muscle relaxants for muscle pain and trauma/injury problems
- hormone replacements—especially testosterone, which helps with pain, mood and energy levels
- addressing nutritional deficiencies

- avoiding MSG and aspartame (both excitatory amino acids that cause an amplification of pain signals)
- avoiding repetitive muscle strains, postural problems, and working to maintain functional range of motion

There have been some encouraging recent studies indicating SAM-e (s-adenosyl-methionine) as well as magnesium can help relieve both the pain and the depression associated with fibromyalgia.

Fibromyalgia and CFIDS are classified as syndromes rather than diseases because they have such a wide range of symptoms.

Other Professional Opinions

William Hennen, Ph.D., a bioorganic chemist from Sandy, Utah, and author of *Fibromyalgia—A Nutritional Approach,* has an interesting take on how the multiple symptoms of fibromyalgia and similar diseases work:

> Fibromyalgia is a syndrome because the identifiable factors vary even though the symptoms are present. This can be likened to the operation of a car. A car will not run if its gas tank is empty. On the other hand, a car with a full tank of gas still will not run if its battery is dead. Thus, a whole list of functions must be present for a car to run properly, while the absence of any one function is sufficient to prevent the car from running.
>
> The human body is infinitely more complex than a car and many more functions must be working properly for the body to operate efficiently. The body, unlike the car, has an interest in its own well-being. The body has many self-healing mechanisms and many self-compensating mechanisms. If given the right tools, sufficient time and correct instructions, the body can heal itself.[2]

An article in the May 2001 *Annals of Internal Medicine* discusses the overlapping symptoms of CFS, fibromyalgia and other syndromes with which we are becoming more and more familiar in this book:

> Unexplained clinical conditions share features, including symptoms—fatigue, pain. Disability out of proportion to physical exami-

nation findings, consistent demonstration of laboratory abnormalities and an association with "stress" and psychosocial factors . . .

Conditions examined were the chronic fatigue syndrome, fibromyalgia, the irritable bowel syndrome, multiple chemical sensitivity, temporal mandibular disorder, tension headaches, interstitial cystitis and the post-concussion syndrome. Many similarities were apparent in case definition and symptoms, and the proportion of patients with one unexplained clinical condition meeting criteria for a second unexplained condition was striking. Overlap between unexplained clinical conditions is substantial.

FMS and CFS clearly affect every part of their victims' lives—not in the least their psychological ability to cope. These are *not* psychological diseases. However, they can have deep-reaching psychological effects when both diagnosis and treatment are so elusive.

One of my favorite authors, Miryam Ehrlich Williamson, a medical and technical journalist from Warwick, Massachusetts, has had FMS since childhood. Actually, she says on her website (www.mwilliamson. com), "I have had fibromyalgia since early childhood, *but fibromyalgia doesn't have me.*"

In her book, *The Fibromyalgia Relief Book: 213 Ideas for Improving Your Quality of Life,* Williamson provides superb coping tools that would apply to people with any chronic disease:

- Reclaim self-esteem and don't let the pain make you more tense. Cultivate self-efficacy. Tell yourself that you *can* do it.
- Build a support system. Surround yourself with people who are positive. Get rid of toxic people.
- Don't expect miracles from your physician and don't overload the physician with too many symptoms. There are many key factors in improvement. Many of them are common sense, but sometimes they're hard to implement.
- You are what you eat, and you should give your body what it needs to heal with proper nutrition.
- Beware of food allergies and sensitivities. Many people have food cravings for carbohydrates and may be allergic to them. Many fibromyalgia patients have problems with malabsorption, and these

may be caused by yeast overgrowth. Because of malabsorption problems, many people with FM may need vitamins and minerals, especially the B complex vitamins, calcium and magnesium.

- Exercise can be tricky. Start out slowly. Increase slowly, but once you start, never skip more than one day at a time.
- When the complexity of FM gets you down, divide and conquer. Pick one problem and focus on it.
- Develop a game plan and keep records.
- Don't give up.
- You're the expert on your body. You know what is best for you. Don't waste energy defending yourself or trying to convince others. Stay focused on your goal and ways to achieve it.[3]

An excellent resource on CFIDS and FMS is Dr. Jacob Teitelbaum's book, *From Fatigued to Fantastic!*[4] and his website, www.endfatigue.com.

HYPOTHYROIDISM

Low thyroid function can be another cause of that "tired all over" feeling.

Hypothyroidism is ten times more common in women than in men and thyroid dysfunction complicates 5–9% of all pregnancies.

Symptoms

The Thyroid Foundation of America notes:

With severe hypothyroidism, you may begin to feel run down, slow, depressed, sluggish, cold, tired and may lose interest in normal daily activities. Other symptoms may include dryness and brittleness of hair, dry and itchy skin, constipation, muscle cramps and increased menstrual flow in women.

There is a simple way to determine if your thyroid function is low:

When you first awaken, before you get out of bed, place an ordinary oral thermometer in your armpit and leave it for 10 minutes. Do this for at least 10 days and, if you are still in your menstrual years, begin on the third day of your period.

Take the reading at the same time every day. If you plan to sleep in, set your alarm to take the reading.

If your reading is consistently below 97.8, it's possible you have a hypothyroid condition. However, when you are ovulating, a normal basal temperature can be as low as 97.0.

Then you can take your results to your doctor and ask for the following thyroid blood tests: TSH, free T3 and free T4 to determine if you have low thyroid function. Sometimes another test for T7, another aspect of thyroid function, can be helpful.

A condition that may have similar symptoms to hypothyroidism called Hashimoto's thyroiditis, is an autoimmune disorder that can be detected by testing for anti-thyroid antibodies.

In Hashimoto's thyroiditis, the immune system develops antibodies against the thyroid. People with Hashimoto's feel tired and this may be interspersed with periods when they feel anxious, unable to sleep and even have tremors. They may have inflammation and enlargement of the thyroid.

If you are found to be thyroid hormone deficient or have Hashimoto's, you'll need prescription medications.

That's the good news about hypothyroidism: It can be diagnosed and there are medications that will address the symptoms, although they won't cure the disease.

Treatment

There are new developments in the treatment of hypothyroidism that your doctor might want to consider.

Three years ago, John V. Dommisse, M.D., a Tucson, Arizona nutritional, metabolic and psychiatric physician, presented a paper to the American Academy of Environmental Medicine that may be the wave of the future in the treatment of this disorder.

For 11 years, Dr. Dommisse said, he has treated hypothyroidism with a combination of Levothyroid, Synthroid and Levoxyl (all containing T4) and Cytomel, which is pure T3.

The Yeast Connection

OK, now where's the yeast connection? We've covered three conditions in this chapter: CFIDS, fibromyalgia and hypothyroidism. They all have overlapping symptoms. Now let's take a look at possible yeast connections for each of them.

CFIDS

Many physicians, including researchers, think viruses cause CFIDS. For a time it was thought that the Epstein-Barr virus was the culprit. Yet, laboratory and other studies showed this was not the case, even though in many people, CFIDS develops suddenly following a viral infection. In other individuals, it seems to come on more gradually.

As I read the reports in the medical and lay literature about CFIDS, it seemed to me that the symptoms in many people with this illness were similar to those experienced by my patients with yeast-related problems.

In the late '80s and early '90s, I attended several conferences on CFIDS and Epstein-Barr. Although the great majority of the speakers focused their attention on viral infections, one CFIDS clinician and researcher, Dr. Carol Jessop, discussed other possible causes of the diseases, including *Candida albicans.*

At CFIDS conferences in San Francisco and Charlotte, North Carolina, Dr. Jessop discussed her findings in working with 1,324 patients with chronic fatigue syndrome she had seen between 1984 and 1990. Her treatment program featured the use of a restricted diet and the antifungal medications Nizoral and Diflucan. At the Charlotte conference, she said she thinks there are multiple causes for these illnesses:

I do think that genetic predisposition and environmental factors such as antibiotics, birth control pills, toxins in the environment and infections all have to be considered.

Eighty-four percent of my CFS patients have recovered to a level where they can remain working 30 to 40 hours a week, and 30% have fully recovered. Yet, 44% of the patients experience some recurrence of their symptoms with premenstrual stress, surgery or other infections.

At another conference, I met Dr. Philip Nelson, a Sarasota, Florida gynecologist, who, along with his wife, Marion, served as professional advisors for a Florida CFIDS group. I saw Dr. Nelson again at a California conference and we began to correspond.

In a letter to me in the spring of 1993, he told me that initially he had been skeptical about the yeast connection, but that he had found that a sugar-free diet and Diflucan helped many women with CFIDS and other problems.

Hundreds of CFIDS patients have benefited from a small article in the CFIDS Association's newsletter in 1998. In a letter to the editor responding to the article, Janet Dauble of Share, Care and Prayer, Inc. of Frazier, California said:

> In our experience with hundreds of people with disabling fatigue, cognitive problems and pain, finding out what foods they have become sensitive to and avoiding them has been the most helpful treatment tool of all.

FIBROMYALGIA

There's not as much solid information about yeast and fibromyalgia, despite the similarity in symptoms many patients experience.

During 2001 and 2002, I've received weekly newsletters from Joseph Mercola, D.O., a Chicago area osteopathic doctor. Here are excerpts from his July 2002 newsletter, "Fibromyalgia Pain is Real—What You Can Do to Relieve It:"

> People with fibromyalgia may experience reductions in their symptoms by eliminating one or more foods from their diet. Following the eating plan seems to help, however, it is quite clear that most people with this disease will not completely relieve their symptoms, even if they follow it perfectly. This is because nearly every person I've seen with fibromyalgia suffers from an underlying emotional component.

HYPOTHYROIDISM

Sometimes treatment for yeast infections should include a hormonal component, especially in women who aren't responding to the anti-yeast regimen.

One theory that has been presented to me, and makes sense, is that hormones are needed to stimulate the immune system in these difficult-to-treat patients. T3 is the number one preferred hormone for this purpose, and the other is DHEA (dehydroepiandrosterone, a steroid hormone, a chemical cousin of testosterone and estrogen that has been shown to have anti-aging, anti-obesity and anti-cancer effects) from the adrenal glands.

Thyroid medication is available by prescription only, but DHEA can be purchased over-the-counter, although it is best to have a baseline measurement of your DHEA level with a blood or saliva test before starting to take it and to discuss it with your doctor.

Judith Lopez's Amazing Story

I was touched by Judith's story of courage and honored when she asked me to write some comments for the jacket of her book, Immune Dysfunction—Winning My Battle Against Toxins, Illness & the Medical Establishment. *Here are excerpts from her story:*

In 1970, at the age of 30, I suddenly became disabled by an illness so bizarre that doctors could not believe it was real. The worst symptom was a crushing, relentless exhaustion. Often I could not even get to the bathroom without collapsing en route. My joints ached, my cognitive abilities were impaired, I developed neurological problems and strange allergies. I saw many doctors. Each one insisted that it was "all in my head."

Some were impatient, some were hostile. One laughed at me. None believed me. But by the mid-1980s, more and more people were developing this "unreal" illness. Our cries for help forced the medical world to give the phenomenon a label—chronic fatigue/immune dysfunction syndrome.

Almost from the beginning, I felt certain that my illness was caused by a damaged immune system. But what had caused it? It couldn't be a contagious bug, for no one ever caught it from me. Unlike a virus, it struck mostly women and did so in a random fashion.

Not until I came across the work of such pioneering doctors as C. Orian Truss, William G. Crook and Carol Jessop did I realize the role of toxic substances in this devastating illness. The cause of my immune system collapse was our chemically altered environment. I was simply being poisoned to death.

My condition deteriorated to the extreme, until I was little more than a skeleton with a heartbeat. I weighed less than 80 pounds and was bedridden, my mouth and throat covered with a growth of *Candida albicans* (fungus). And then, when it was almost too late, I came under the care of a doctor who understood my illness. Dr. Vincent Marinkovich saved my life when the task seemed all but impossible. Under his guidance, I removed the major toxins from my environment, took antifungal drugs and supplements, and learned to eat a healthy, immune-supporting diet.

Little by little, I felt my body reconnecting to itself. And I took the first tentative steps back to health, re-entering at last the world and the life that I thought I had lost forever. My experience has convinced me that CFS need not be mysterious or frightening. Once the underlying toxic causes are understood, the pathway back to health becomes clear and is available to everyone.[5]

Marianne Hauwelaert's Story

Hello Dr. Crook,

My name is Marianne. I am 37 years old. I am happily married for 17 years and have two wonderful children. We live in a nice house with a large (to Belgian standards) garden. All this sounds nice indeed, but there's a dark side to my story.

For almost eight years now, I've suffered from poor health. The first four years I was told that all my complaints were caused by depression, and my doctors wanted me to take antidepressants. I wasn't convinced and refused to take them. So, I was classified as a "difficult patient." After four years, I got the diagnosis of Chronic Fatigue Syndrome.

In the meantime, my list of complaints was getting longer by the day. I was now at the stage that I couldn't get up from my chair or lift a cup of tea. Again they prescribed a number of drugs to ease my symptoms.

I refused because I felt I could not take drugs if the doctors couldn't even tell me what was wrong with me. It didn't feel right for me to fill my body with more chemical pollutants because it was already sick enough.

Two years passed. I was just a zombie, who was trying to find my way in this terrible world. I had lost my confidence in doctors and was trying to cope on my own. Luckily, I had some very supportive people behind me. Otherwise, I probably would have taken my own life. During that time, I had noticed that when I ate certain things, my symptoms got worse.

I searched the Internet and read everything I could find about chronic fatigue syndrome. Then, one day, I found a connection with *Candida albicans.* I ran into an article by Dr. Elmer Cranton, and I knew that candida was the right track to follow. Sadly enough, there was no doctor that would take it seriously. Can you believe this? What a bunch of . . . . Sorry, it just makes me feel so angry.

I had to play my own doctor. I bought all your books, via Internet, because they aren't available here. Well, dear Doctor, after eight years, I feel like I've been born again. Six months ago, I started the diet and, later on, as I read your book, *Yeast Connection Success Stories,* I started to take Diflucan.

I also take caprylic acid, flaxseed oil, extra vitamins, lots of good bacteria and something to clean up the liver. I hope to be able to discontinue the Diflucan.

The results are unbelievable. Now I am even more angry with these so-called "doctors."

I am so grateful to you. It is all so true, but then again really sad that most doctors aren't ready to see yet. I guess it will take years before your message gets here. Meanwhile, I'll play one of your disciples and spread the message. Many, many thanks. I really hope that this message reaches you personally, and maybe, when you find the time, you could send me a little word back.

Maybe some day, as I regain my health, I'll translate your book into Dutch. Oh, that would be something!

Kindest Regards,
Marianne Hauwelaert
Belgium

MY COMMENTS

Depending on their background and training, professionals interested in chronic fatigue emphasize different treatment approaches, and no one claims to have all the answers.

Certainly, I do not.

Yet, in my practice, I've found that many people with chronic fatigue, fibromyalgia, headaches, muscle aches, memory loss, digestive disorders and other symptoms can be helped when they follow a comprehensive treatment program that includes:

- dietary changes
- prescription and nonprescription antifungal medications
- avoiding foods that cause sensitivity reactions
- getting rid of pollutants in their homes
- taking nutritional supplements, including probiotics
- testing thyroid function and taking thyroid supplements (if needed)
- receiving psychological support

Based on my own experiences in practice, and those of other physicians, including Drs. Jessop and Nelson, I feel that CFS/CFIDS and fibromyalgia are often yeast-related. People with these disorders seem to develop them because their immune systems are disturbed. When their immune systems are weakened, viruses are activated, yeasts multiply, food and chemical allergies become activated and nutritional deficiencies develop in a leaky gut environment.

I'm not claiming that the common yeast, *Candida albicans,* is the cause of CFS/CFIDS/fibromyalgia and other disorders that affect many people. Yet, increasing evidence shows that a sugar-free special diet, probiotics and antifungal medications may help many people with these chronic health disorders get well.

Because of my success in treating and helping many patients using this program, I've emphasized the "yeast connection" to health problems that affect millions of women. I realize that I may be one of the few doctors who sees the whole picture.

As pointed out by John Godfrey Saxe in his classic poem, each blind man described the elephant according to what he had seen, and each blind man had a different story. One of the men looked only at the side of the elephant, another felt his trunk and another felt his sharp tusks. One man felt his knee and said, "'Tis clear enough, the elephant is very much like a tree."

Another man felt his ear and said, "This marvel of an elephant is very much like a fan," and yet another man, after seizing on the swinging tail said, "I see, the elephant is very much like a rope."

RESOURCES

- CFIDS Association of America
  Website: www.cfids.org
  Charlotte, NC
  Voice Mail: (800) 442-3437
  Resource Line: (704) 365-2343

- American Fibromyalgia Syndrome Association
  Tucson, AZ
  Website: www.afsafund.org
  (520) 733-1570

- Immune Support.com website, includes information about both chronic fatigue and fibromyalgia:
  www.immunesupport.com

- Thyroid Foundation of America
  Website: www.tsh.org
  (800) 832-8321

- Website for thyroid expert Mary Shomon
  www.thyroid-info.com

- Dr. Joseph Mercola's website:
  http://mercola.com

- Dr. John Dommisse's website
  www.johndommissemd.com

Shames, Richard. *Thyroid Power.* HarperResource, 2002.

Shomon, Mary. *Living Well With Hypothyroidism,* HarperResource, 2000.

REFERENCES

1. Goldenberg, D. *Fibromyalgia: A Leading Expert's Guide to Understanding and Getting Relief from the Pain That Won't Go Away.* Perigee, 2002.

2. Hennen, William, *Fibromyalgia—A Nutritional Approach.* Woodland Publishing, 1999.

3. Williamson, Miryam. *Managing Fibromyalgia: A Nonmedical Approach—A patient's guide to do-it-yourself health care.* Royal Publications Inc., 1999.

4. Teitelbaum, Jacob. *From Fatigued to Fantastic!,* Avery, 2001.

5. Lopez, Judith. *Immune Dysfunction—Winning My Battle Against Toxins, Illness & the Medical Establishment.* Millpond Press, 2001.

CHAPTER 18

Interstitial Cystitis

During my many years of pediatric practice, I saw and treated many patients—especially young girls—with urinary tract infections. Often these infections would respond to a short course of an antibacterial drug. Yet, when children continued to be troubled by recurrent infections, I would refer them to a urologist for examination and further therapy.

After becoming interested in adults with complex health problems, I began to receive letters and phone calls from women who were troubled by repeated bladder problems.

The occasional bout of cystitis or a bacterial bladder infection can usually be cleared up with a course of antibiotics with few long-term consequences.

Interstitial cystitis is a quite different condition: It's a painful, chronic inflammation of the bladder. Its origin is unknown, but we do know that it's not caused by bacteria. We also know it is not caused by stress, and it is not a psychosomatic disorder.

Patients with severe IC may experience a need to urinate as many as 60 times a day. That frequency may be the only symptom, or others may develop:

1. Urgency—The need to urinate immediately, sometimes accompanied by pain, pressure or spasms.
2. Pain—This may be lower abdominal pain or discomfort in the urethral or vaginal areas. There is often pain with intercourse.
3. Women's symptoms often worsen during menstruation.

It's primarily a women's problem. Of an estimated 700,000 chronic sufferers of IC in the United States, approximately 90% are women.

Interestingly, a recent study published in the *Journal of Urology* says that 10% of women with IC also have vulvodynia.

Of even more interest: In 1997, researchers at Temple University in Philadelphia found that people with interstitial cystititis often had other chronic conditions:

- more than 40% had allergies
- 25% had irritable bowel syndrome
- 22% had sensitive skin
- a significant percentage had the following chronic conditions:
  —fibromyalgia
  —migraine
  —endometriosis
  —chronic fatigue and immune deficiency syndrome
  —asthma

Plus, they were:

- 100 times more likely than the general population to have inflammatory bowel disease
- 30 times more likely to have systemic lupus erythematosus (an autoimmune disease)

Does any of this sound familiar?

This suggests to me that IC may be part of the universe of symptoms of chronic yeast overgrowth.

I spoke with Phillip Mosbaugh, M.D., an Indianapolis urologist who has for years used an anti-yeast regimen to treat some of his patients who have not responded to conventional IC therapy.

He said, "I think there's a subset of patients that have the diagnosis of IC along with fibromyalgia, irritable bowel syndrome, chronic fatigue syndrome, migraines, chemical allergies, etc., and candida is a part of that. Not everybody gets better, but some did, and some did very well."

The Interstitial Cystitis Foundation's (ICF) newsletter underscores the severity of the problem for its victims:

Epidemiological studies reveal that:

- It takes on average 5–7 years to get diagnosed, and sometimes even longer.
- The quality of life of IC patients has been shown to be worse than that of patients undergoing dialysis for end-stage renal disease.
- Economic impact is estimated to be $1.7 billion per year when combining medical expenses and lost wages due to inability to work.

The ICF adds: "Suicides occur every year because patients are left in severe pain with nowhere to turn for help. Because standard urologic tests are negative and physicians are often not familiar with the condition, patients are often told that their symptoms are "all in their heads," or that the symptoms are caused by stress, thereby minimizing or invalidating the patient and compounding an already devastating condition."

DIAGNOSIS

The National Institute of Diabetes, Digestive and Kidney Diseases (NIDDK) of the National Institutes of Health says since there is no test to definitively diagnose IC, it is usually diagnosed by ruling out other urological problems. These can include urinary tract or vaginal infections, bladder cancer, bladder inflammation or infection caused by radiation to the pelvic area, other types of cystitis, kidney stones, endometriosis, neurological disorders, sexually transmitted diseases, low-count bacteria in the urine and, in men, prostatitis.

These other conditions can sometimes be identified by a variety of laboratory tests.

Frequently, cystoscopy (the insertion of an instrument to look inside the bladder) under anesthesia is necessary. This test can give an idea of bladder capacity and bladder wall inflammation or other abnormalities of the bladder wall that may help with a diagnosis.

TREATMENT

There is no complete cure for IC. The NIDDK says symptoms may disappear without explanation or respond to a change in diet or treatment. Symptoms may even disappear without explanation and may return equally without explanation days, weeks, months or even years later.

The NIDDK says treatment options include:

- Over-the-counter medications: Aspirin and ibuprofen that may help relieve pain.
- Oral drugs: Elmiron (pentosan plysulfate sodium) has been shown to improve symptoms in 38% of patients treated with it. It may take at least six months of therapy to determine if Elmiron will be helpful.
- Other oral medications, such as antidepressants and antihistamines, which are helpful to some patients.
- Bladder distention: This is similar to a cystoscopy, which helps in some patients, perhaps because it increases the capacity of the bladder and may interrupt pain signals.
- Bladder instillation: Also called a bladder wash or bath, in which a catheter is inserted into the bladder and the bladder is filled with a wash or solution, most frequently DMSO (dimethyl sulfoxide), and left there for 15 minutes before it is expelled.
- Transcutaneous Electrical Nerve Stimulation (TENS): Mild electric pulses enter the body for a predetermined period of time at regular intervals, either through wires placed in the lower back, just above the pubic area, or in the vagina or rectum. It may increase blood flow to the bladder, strengthen pelvic muscles that help control the bladder or trigger the release of natural chemicals that block pain.
- Diet: Acidic foods may irritate the condition, and many patients find that eliminating them is helpful. Others have found that artificial sweeteners are irritants.
- Exercise: Gentle stretching exercise, such as yoga, may help.
- Surgery: This is a last resort that may not relieve symptoms. It involves cauterizing ulcers in the bladder (if they are present) or en-

larging the bladder using a section of the bowel or, rarely, even removing the bladder entirely.

IC AND THE YEAST CONNECTION

In the mid-1980s, at a conference in the UK, I met Angela Kilmartin, an English woman who published a book in the States entitled *Cystitis—The Complete Self-Help Guide*. She warns against using antibiotics as a cure all for bladder problems:

> Antibiotics are the commonest medical way of dealing with bladder troubles . . . The action of antibiotics is to remove bacteria—all bacteria, not just the bad, but also the good . . . On a prolonged course of antibiotics lasting several weeks (which the medical profession is still fond of prescribing in cases of recurrent cystitis) . . . the patient feels washed out and lethargic and finds it immensely difficult to get out of bed.[1]

In her continuing discussion, Kilmartin points out that antibiotics may be needed for women with severe urinary tract infections.

In the early 1990s, I read Dr. Larrian Gillespie's book, *You Don't Have to Live With Cystitis.*

In her comprehensive discussion of cystitis in women, Dr. Gillespie emphasized the importance of urine cultures prior to the use of antibiotics. She also sharply criticized the practice of prescribing antibiotics over the phone because they can cause other problems.

In November 1992, at a candida seminar sponsored by an Indiana group, I met Terry Oldham, who told me about her favorable response to a comprehensive treatment program that featured Diflucan and a sugar-free diet. Because Terry's interstitial cystitis had improved dramatically on Diflucan and diet, I asked her to write me. Here are excerpts from her six-page letter.

Terry Oldham's Story

I had lived with interstitial cystitis for 12 years, and it is much more serious and involved than a bacterial infection of the bladder. I developed tiny hemorrhages in the lining of my bladder that allowed urine to leak through the wall of my bladder; this, in turn, caused constant severe burning and pain.

I was diagnosed with IC in 1981. My symptoms at that time were urinary frequency, burning, pelvic pain, bladder spasms, fatigue and mild to moderate digestive problems.

During the next four years, Terry's symptoms continued and gradually worsened. And by 1989, because of severe bladder pain, frequency, pelvic pain, bladder spasms, fatigue, muscle and joint pains and chemical sensitivities, she had to give up her position as staff nurse at the Methodist Hospital in Indianapolis.

My pain and fatigue were so severe I could barely stand up for more than an hour at a time. I had to urinate every 20 minutes on a bad day, and at least once an hour on a so-called "good" day . . .

For the next two years, Terry tried every treatment available for IC, and none of them relieved her pain.

Then, in June 1991, after she learned about Diflucan, she sought the help of her personal physician, Stephen Heeger, D.O. He prescribed Diflucan, which she took in gradually increasing doses until she was able to tolerate 100 mg. a day. By November 1991, just five months later, her bladder pain and frequency were 50% improved.

Her problems were further relived by a dietician who had worked with other IC patients who had multiple food sensitivities and related problems.

To relieve her remaining symptoms, Terry's physician gave her intramuscular injections of vitamin B_{12}, supplemental vitamins and minerals, flaxseed oil, probiotics and CoQ_{10}. She also continued to take Diflucan intermittently, and she said:

The fatigue I was having has lessened quite a bit since adding these new things to my program. My bladder pain and frequency are still 90% improved most of the time, and I feel I have a quality of life that would not have been possible if I had not followed a comprehensive treatment program.

MY COMMENTS

Terry Oldham has been my friend since she first came up to me at the conclusion of my presentation at Hoosiers for Health, a conference sponsored by an Indianapolis health food store.

Terry's story is an important one for many reasons, including:

- It shows that IC, like many other health problems that trouble women, are interrelated and as Indianapolis urologist Phillip Mosbaugh, M.D., and others have pointed out—it's not just a disorder of the bladder.
- It shows that with courage, persistence, prayer, faith and a variety of therapeutic measures, people with IC and other often devastating illnesses can regain their health and their lives.

A CLINICAL STUDY

Possible Yeast Influences in Interstitial Cystitis: A Study* by Phillip G. Mosbaugh, M.D.

During the past decade, Dr. Mosbaugh has treated more than 500 IC patients in his urology practice in Indianapolis and served as medical adviser for the Indiana Chapter of the Interstitial Cystitis Association. He followed the treatment of Terry Oldham. He said:

People with interstitial cystitis have multiple symptoms with multiple causes, and multiple therapies are necessary. Also, everybody is different. This makes it difficult to do a single study. A lot of research is going on today, and a number of medical researchers and clinicians are beginning to look at IC as a total body problem and not just one affecting the bladder.

- The International Health Foundation, Inc. supported this study with funds obtained from the Pfizer Corporation. Additional support was provided by Allergy Resources, Inc., Antibody Assay Laboratory, Great Plains Laboratory and Klaire Laboratories, Inc.

Dr. Mosbaugh's clinical study of IC patients in 1997 and 1998 began with the premise:

> The purpose of this study is to explore possible yeast influences affecting the patient's immune system and subsequently resulting in a variety of symptoms, including those leading to the diagnosis of interstitial cystitis. From patient surveys, it is apparent that up to 35% to 40% of IC patients may suffer not only the typical bladder symptoms but also from other problems, including irritable bowel disease, chronic fatigue syndrome, fibromyalgia, headaches, vulvodynia, muscle and joint pain, irritability and allergic states . . .

Each of the 15 IC patients selected for the study gave a history of prolonged antibiotic use, an elevated score on the Yeast Questionnaire (greater than 250 total points), failed conventional IC therapy and prior diagnosis based on standard criteria.

Upon initiation of the study, the patients followed a rigid antifungal diet for six weeks. They also took Omega-3 fatty acids (one tablespoon of flaxseed oil daily), as well as vitamin/mineral complex supplements each day during the duration of the study. After four weeks on the rigid antifungal diet, the antifungal medication, Diflucan, was administered in an initial dose of 400 mgs., followed by 200 mgs., daily for four months. One week after starting Diflucan, a probiotic was begun as a nutritional supplement.

In a February 1998 report on his study, Dr. Mosbaugh said:

> There are five patients in this study who have clearly improved, and a sixth patient who was improving and became pregnant. She also had improved. I talked to her a couple of times, and she attributed her improvement to the antifungal medication and diet. These patients felt better, lost weight, vaginal discharge and fibromyalgia symptoms improved, vaginal itching got better.
>
> Another patient with fibromyalgia, irritable bowel and headaches, also got better. These five women all felt there was a definite impact from the treatment program.

You have to realize these people had a lot of things on their plate—bowel problems, joint aches and pains, fatigue, headache and other symptoms. It was a tough group of patients. If we do the study again, maybe we'll look at some of the patients who weren't the worst ones . . .

I spoke with Dr. Mosbaugh in March 2002, and he had some more thoughts on the problems of patients with IC that seems to be yeast-connected:

These patients have weakened immune systems—they've gone to the bank one too many times—and they can just never catch up. But the patients who really stick with the comprehensive program can be helped. I've seen several dramatic turn-arounds—not just with their bladder symptoms but everything.

Dr. Mosbaugh also provided some hope that there are changing attitudes in the medical establishment, or at least some parts of it:

At some of our national meetings, the leaders in IC are now making comments that some of these patients have a total body problem. Although this may sound like a small statement, I really feel like it's progress, because not too many years ago they were looking at these patient's bladders and not looking at the whole patient.

MY COMMENTS

In my opinion, interstitial cystitis develops from many different causes, just as do many and most of the health problems that I've discussed in this book. Here are a few of them: endometriosis, vulvodynia, chronic fatigue, PMS, depression and sexual dysfunction.

As pointed out by Dr. Mosbaugh, women with IC, almost without exception, are troubled by symptoms in many other parts of the body. The bottom line: IC is more than a disease of the urinary bladder.

Dr. Mosbaugh's study, along with reports I received from women with interstitial cystitis, shows clearly that:

a) In many women, IC is yeast-connected and long-term antifungal therapy may help.
b) An appropriate diet is essential.
c) Nutritional supplements are important if the person is to obtain optimal results.
d) Determination and persistence by both physician and patient are necessary if the treatment program is to succeed.

RESOURCES

The Interstitial Cystitis Association has information, support and the latest research. Website: www.ichelp.com, (800) HELP ICA

Moldwin, R.M., *The Interstitial Cystitis Survival Guide—Your Guide to the Latest Treatment Options and Coping Strategies,* New Harbinger Publications, Inc., 2000.
Gillespie, L., *You Don't Have to Live with Cystitis,* Avon Books, 1996 (updated edition)
Simone, C., *To Wake in Tears: Understanding Interstitial Cystitis,* IC Hope Ltd., 1998

REFERENCE

1. Kilmartin, A., *Cystitis—The Complete Self-help Guide,* Warner Books, New York, 1980. (Out of print, but available on special order from Amazon.com.)

CHAPTER 19

Chemical and Mold Sensitivities

During my childhood in the 1920s, my parents, siblings and I lived on a three-acre "mini" farm, along with one or two cows and 100 chickens. Scattered over our home place were many oak trees, and my mother enjoyed growing vegetables and flowers. Compost piles made from decaying leaves and cow manure provided nutrients for these plants. My mother grew so many vegetables that I earned pocket money by selling them to the neighbors.

Neighbors and friends would say, "Mrs. Crook has a green thumb. Her vegetables grow better than anybody else's." I can't recall my mother spraying insecticides, weed killers or other chemicals in our garden.

I walked to school a mile away. I skated, rode my pony or my bicycle after school, and during the summer, I played tennis, baseball and other sports. (We had a tennis court and a small baseball "diamond" on our back lot.) In addition, I had to do chores like cutting the grass and raking leaves, so I spent a lot of time outdoors.

Few automobiles came down the road to pollute the air with lead and other toxic substances contained in automobile exhaust. Houses were made out of wood or brick and heated by radiators and wood fires. Clothing and home furnishings were made from natural fabrics that didn't put out offensive odors. When I was growing up, we ate good food and breathed clean air.

I don't want life to sound perfect, because it wasn't. Family members, friends and neighbors experienced health problems almost unheard of today. One of my sisters almost died with typhoid fever, and a brother almost died with pneumonia. Scarlet fever, measles and other childhood diseases caused death and disability in many people. So did diphtheria, "lock jaw" (tetanus) and polio.

During the past 75 years, amazing, fantastic and breathtaking changes have taken place. Because of advances in medicine, many lives have been saved and much suffering has been prevented or relieved. Although women in the 21st century are living longer than their grandmothers and great-grandmothers, they're developing more chronic complaints, including vaginitis, PMS, fatigue, endometriosis, depression, digestive disorders, respiratory problems, urinary problems, weight problems, food and chemical sensitivities, osteoporosis, cancer and other problems.

CAUSES OF CHEMICAL AND MOLD SENSITIVITIES

Why are people developing these problems? Here are some of the reasons:

1. nutritionally deficient diets loaded with sugar, white flour and unhealthy fats
2. they're taking in antibiotics—not only those prescribed by their physicians—but antibiotics in animal foods and dairy products
3. prescription drugs of many types
4. chemicals of all sorts inside and outside our homes

Today our homes, apartments, farms, schools and public buildings are contaminated by pesticides and toxic chemicals, and many people are developing allergies to house dust mites and common molds. And some people are affected by poisonous black molds, including *Stachybotrys chartarum,* commonly found on chronically damp walls in homes and buildings.

If your health problems are "yeast-connected," even though you change your diet, take antifungal medication and nutritional supplements, you may continue to experience problems because chemicals and molds in your home can be two of the biggest contributors to your health problems.

Chemicals

Hazardous and potentially hazardous chemicals have been around for decades—even centuries. After the Great Depression in the 1930s and after World War II in the 1940s, Americans fell in love with chemicals. More and more were developed each year and were hailed as signs of "progress." Here are a few of them:

- Lead and other additives were added to the gasoline of our growing number of automobiles because they helped the cars run better.
- DDT and other insecticides were sprayed in our homes and other places because people felt they helped get rid of "pesky" insects.
- Weed killers and insecticides were sprayed on the soil and crops of our farms because these chemicals seemed to help farmers raise more food to feed our growing population and export to other countries.
- Protective coatings were added to upholstered furniture and carpets to prevent stains and simplify cleanup.

Courageous Pioneers Began to Warn Us

In 1962, in her classic book *Silent Spring,* Rachel Carson compared the threat of environmental destruction from pesticides and other environmental chemicals with the risk of nuclear war. Most Americans, including legislators, farmers and food processors, paid little attention. About the same time, Theron Randolph, M.D., a Chicago internist, allergist and specialist in environmental medicine, warned:

> While it is true that outdoor air pollution is a significant source of exposure, a far greater threat is posed by the presence of indoor . . . air pollution . . . Many household products give off noxious fumes. Indoor air pollution is particularly dangerous because exposure is so constant.[1]

Recent books on the topic include Theo Colburn's book *Our Stolen Future,* and *Living Downstream* by Sandra Steingraber.

During the past 30 years, Dr. Randolph and several hundred physicians and other professionals in the American Academy of Environmental Medicine (AAEM) have been studying and treating patients with chronic health disorders related to chemical overload. They have found that by lightening this load, many patients with puzzling and persistent health problems improve.

People with environmentally induced illness (EI) or multiple chemical sensitivity (MCS) and their physicians have faced much of the same sort of skepticism and hostility as those who have been working to bring "the yeast connection" into the medical mainstream. Yet, happily, the situation seems to be changing.

After graduating from medical school in the late 1950s and preparing for a career as a cardiac surgeon, the personal and professional life of William Rea, M.D., was forever changed. Dr. Rea, a Dallas surgeon who later became board-certified in environmental medicine, had his home sprayed with insecticides, which made him and other members of his family sick.

Over the ensuing decades, Dr. Rea has published a number of scientific articles and written several books that document the role that chemical exposures play in adversely affecting many parts of the immune system. And he said, "Your resistance resembles a rain barrel, and chemicals in your environment are like pipes draining into the barrel. When you're exposed to many chemicals, your barrel overflows and you develop symptoms."[2] For more information see www.ehcd.com.

In the early '80s, Sherry Rogers, M.D., a Syracuse, New York physician, began writing and publishing scientific articles and books about allergies and food and chemical sensitivities. Included in her many publications is her 1998 book, *Depression: Cured at Last!* Here's an excerpt from this book:

> If you think you have trouble believing that everyday chemicals found in the average home and office can make people depressed, you can imagine how I felt . . . One of the big problems with chemical sensitivity is that it usually does not happen dramatically or quickly, but rather insidiously: It sneaks up on someone . . . Furthermore, most

chemical sensitivity sneaks up on a person over a period of weeks, months or years, as the detoxification system is slowly damaged.[3]

During the past two decades, two "mainstream" professionals, Nicholas A. Ashford, Ph.D., J.D., and Claudia S. Miller, M.D., M.S., have discussed the role that chemical pollutants play in making people sick.

And in the second edition of their book, *Chemical Exposures: Low Levels and High Stakes,* they cite a study that compares the symptoms of patients with multiple chemical sensitivities (MCS) with the symptoms of patients with Chronic Fatigue Syndrome (CFS) and fibromyalgia (FM).[4]

The Hormone Connection

I learned a great deal about the adverse effects of chemicals on hormones from Mary Lou Ballweg, president of the Endometriosis Association. She made me aware that PCBs (polychlorinated biphenyls used in industry until they were banned in the 1970s) and other toxins were playing a part in causing endometriosis and other endocrine and immune problems.

We've already talked about the high rate of endometriosis in monkeys exposed to minute amounts of dioxin, a byproduct of industrial processes that involve chlorine, including paper and plastic manufacturing.

In his book, *Hormonal Chaos,* Sheldon Krimsky, Ph.D., a professor in the Department of Urban and Environmental Policy at Tufts University, discusses the hazards of environmental chemicals. He puts forth the environmental endocrine hypothesis, the assertion that chemicals called 'endocrine disruptors' are interfering with the normal functioning of hormones in animals and humans.[5]

He added, "One-third of these patients have been found to have low T-cells (cells that power your immune system) . . . It has become increasingly apparent that these chemical sensitivities are disappearing and the T-cells are returning to normal (following treatment with nystatin . . . indicating that the low T-cell counts were caused by *Candida albicans*."[6]

Dr. D. Lindsey Berkson, a consulting scholar at the Center for Bioenvironmental Research at Tulane and Xavier universities in New Orleans, is the author of *Hormone Deception,* which explores the link of pollutants that mimic hormones and various illnesses in children, women and men. Berkson calls pollution and its widespread effects on human health, "an intimate dance." She says, "We affect the environment with our technological culture and, in turn, endocrine-disrupting chemicals have the potential to affect us all. We could call it the shadow side of technology."

Berkson thinks *all* illness must now be at least investigated as having a possible link with our environment. This led her to write her next book, *Natural Answers for Women's Health Questions,* in which she explores 220 illnesses, from endometriosis to breast cancer, with their possible connection to environmental hormonal pollutants, along with human hormonal and nutritional deficiencies, and tells us how to work with complementary medicine to unravel these problems.

She adds, "Many patients with chronic illnesses have chemical sensitivities. We don't know exactly why, but Dr. Truss says in research dating back to 1978 that there is a subgroup of patients with candidiasis 'with severe intolerance to virtually all chemicals.'"

Based on this information, I was able to help many of my chemically sensitive patients by adding antifungal medications to their treatment program.

Molds

Molds inside spacious homes, small and large apartments, schools, offices, factories (and other workplaces) are making people sick, and they may be contributing to the "sick all over" feeling that so many people with yeast overgrowth experience.

In his cover story article in *USA Weekend* (July 19–21, 2002), "When Mold Takes Hold," Arnold Mann provides clear—even frightening—documentation. Here are excerpts:

> Since 1999, when *USA Weekend* magazine first published the story of a mold-stricken family in Dripping Springs, Texas, reports of mold-related illnesses and insurance claims have skyrocketed from California and Texas to Louisiana and New York. Families have abandoned mold-plagued houses. Affected schools have closed and relocated

children. Insurance companies hit with mounting claims for moldy homes have raised premiums and, in some regions, stopped selling homeowner policies altogether.

Mann said, "Occasional growth of common molds, like *Cladosporium* and *Alternaria,* rarely pose a significant health threat. But when a leak goes untended and timbers or wallboards become saturated, it doesn't take long—a few weeks, perhaps, for mold to grow and fill the air with spores."

He also cited reports from researchers who found that people in many parts of the United States who live in homes that show water damage develop respiratory symptoms because of mold overgrowth.

Recent studies suggest the same problems exist in apartment buildings. Mann reports: "A 1999 Mayo Clinic study pegged nearly all of the chronic sinus infections afflicting 37 million Americans to molds . . ."

When toxic molds such *as Stachybotrys, Aspergillus, Versicolor* and some species of *Penicillium* are involved, it's another matter entirely. These molds—which grow in damp, dark places and often are hidden behind walls, under floors and above ceilings—produce dangerous airborne mycotoxins.

"Many doctors believe they cause a raft of serious ills, including flu-like symptoms, chronic fatigue, memory impairment, dizziness and bleeding of the nose and lungs, while others say the science isn't there yet to make that claim," says Mann.

Jeffrey May also includes references to molds in his book, *My House Is Killing Me!* And he said, "People who breathe in mold spores may develop allergic reactions that may make them cough, sneeze and show other respiratory symptoms, but mold can bother people who do not have mold allergy at all."

And he said that the black mold *Stachybotrys chartarum* so contaminated a Texas home "that the family had to abandon the property and bulldoze the house. The husband had experienced memory loss and the child became asthmatic."[6]

***Indoor Molds Are Causing Problems for More and More People*—**
During my early years of allergy practice, I learned that allergic reactions to *Alternaria* and *Hormoderum* and other indoor molds caused na-

sal congestion, sinusitis, bronchitis, asthma and other symptoms in a number of my patients. During a visit to my office in the late 1980s, one of my adult patients with food and chemical sensitivities and yeast-related problems said:

> I've just returned from visiting my daughter and my grand-child. They live in an old house that smells musty and moldy. Almost from the moment that I set foot in the door, I began to "feel funny." Although my nose became stuffy, my worst symptoms were feeling tired and in a fog. These symptoms worsened the longer I stayed in my daughter's home. When I returned to my home and breathed fresh air, my symptoms began to subside. I told my daughter:
>
> When I come back for our next visit, we'll get together at my motel or anywhere but your beautiful old home.

In his book, *Tired All the Time,* Dr. Ronald Hoffman connects mold and these types of symptoms. He tells the story of a 22-year-old woman who was suffering from extraordinary fatigue, headaches, jitteriness and inability to concentrate. When her physician could not find a physical cause for her symptoms, he referred her to a psychologist who concluded that she was having a severe emotional problem.

Then, she went to see Dr. Hoffman, who took her history and learned that her symptoms developed when she moved to a new building. So, he instantly suspected that she was suffering from the "sick building syndrome." He asked her to take mold plates to her office and expose them. When the mold plates were returned and analyzed, they showed a total fungus and mold growth too numerous to count. The woman's employer hired a firm to clean the ducts in the building's cooling system, and her fatigue and other symptoms disappeared.

Almost a decade earlier, environmental allergy pioneer, Dr. Sherry Rogers, published three studies describing the role of mold in making people sick and what could be done to identify these molds and help people with these problems.

Then as I was gathering information for this book, I read again Arnold Mann's cover story in the August 18–20, 2000 issue of *USA Weekend* entitled, "Mold in Schools: A Health Alert." This article told how

molds in schools were making children and teachers s
caused by molds included asthma, sinus infecti
coughing, eye and throat irritations, chronic fatigue a...
pairment. The mold problem was so bad that El Paso, Texas n...
$4.2 million for mold renovations in 14 schools.

According to Mann, mycotoxins caused by airborne toxins can
cause even more serious problems, including chronic fatigue, loss of
balance and memory, irritability and difficulty speaking.

MY COMMENTS

I can see how my discussion of chemicals can be frightening—even
overwhelming. Yet, by becoming more aware of the health problems
they cause, you'll be able to take steps to avoid or control them.

If you're troubled by vaginitis, endometriosis, fatigue, fibromyalgia
or other yeast-connected problems I've discussed in this book, you already know there's no "quick fix." Yet, there are things you can and
should do.

Here are a few of my suggestions, especially if you're troubled by
sensitivities to perfume and other odors, or if you're troubled by respiratory problems, including sinusitis or asthma:

1. Get copies of two superb books that focus on indoor chemicals—*Staying Well in a Toxic World* and its supplement, *Staying
   Well in a Toxic World: A New Millennium Update*—by Lynn Law-
   son. These books are not only authoritative and carefully documented, they are as readable as paperback novels.

2. Don't smoke, and don't let people smoke in your home. People
   in homes where others are smoking experience twice as many respiratory infections as individuals in smoke-free homes, and
   these infections set up a vicious cycle of other health problems.

3. Don't spray insecticides inside or outside your home, and keep
   your windows closed on the days your neighbors spray their
   houses.

4. If your office is making you sick, bring information to your employer or seek a job in a less-polluted environment.

5. Get a copy of Jeffrey C. May's 338-page book published in 2001, *My House Is Killing Me!—The Home Guide for Families with Allergies and Asthma*. May explains how air conditioning, finished and unfinished basements and other home features affect air quality. He offers a step-by-step approach to identifying, controlling and often eliminating the sources of indoor air pollutants and allergens. May also discusses many other things in your home that can bother you. Among them: "off-gassing" of chemicals like formaldehyde from furniture and carpets; old, buried gas or oil tanks or by-products of a manufacturing operation that may have leached toxic chemicals into the soil and even into your basement.

The use of chemicals or pesticides in your yard or those of your neighbors can also have serious effects on you.

REFERENCES

1. Randolph, T. G., *Human Ecology and Susceptibility to the Chemical Environment,* Charles C. Thomas, Springfield, IL, 6th printing, 1978.

2. Rea, W. J., *Chemical Sensitivity,* Vol. I–IV, CRC Press, Boca Raton, FL, 1992–1998.

3. Rogers, S., *Depression: Cured at Last!,* Prestige Publishing, Syracuse, NY, 1998 or www.prestigepublishing.com.

4. Ashford, N.A. and Miller, C.S., *Chemical Exposures, Low Levels and High Stakes,* Van Nostrand Reinhold, New York, 1991, p. 70.

5. Krimsky, S., *Hormonal Chaos—The Scientific and Social Origins of the Environmental Endocrine Hypothesis,* The Johns Hopkins University Press, Baltimore, 2000.

6. May, J. C., *My House Is Killing Me—The home guide for families with allergies and asthma,* Johns Hopkins University Press, Baltimore, MD, 2001, pp. 18–19, 272.

7. Berkson, Lindsay, *Hormone Deception,* McGraw-Hill Contemporary Books, 2001.

CHAPTER **20**

Infertility

One American couple out of every six is troubled by infertility. This translates to about 10% of the adult population of reproductive age, some 6 million people.

The causes are multiple and complex. Some causes of infertility are known: a malfunction in ovulation, endometriosis, polycystic ovary syndrome, irregular periods, fibroids, pelvic inflammatory disease, environmental toxins and sperm problems.

Yet, others are unknown, and a woman may fail to conceive even when sperm are active and normal, and a comprehensive infertility investigation comes up empty-handed.

Endometriosis is one of the primary causes of infertility in women, and since we know that many women with yeast-related health problems have endometriosis, it's a leap—but perhaps not too big a leap—to suggest that *Candida albicans* might have a connection to infertility.

Should infertile couples who want to conceive consider investigating candidiasis as an underlying factor in their infertility? I'll say a cautious, "Why not?" It's certainly worth investigating if you can't find a reason for your infertility. It's even more likely if you have other symptoms of dysbiosis. I'll go as far to say it becomes even more likely if you have been diagnosed with candida-related health problems.

OTHER DOCTORS' OPINIONS

I have spoken with many colleagues who agree with the theory that infertility is somehow connected to candida yeast overgrowth.

Among them is Dr. Charles Resseger, D.O., of Norwalk Ohio:

I've treated dozens of patients with an infertility problem who, following a comprehensive treatment program which included anti-candida therapy, became pregnant. What I do with these patients is to tell them, "Don't get yourself pregnant until I get you squared around." If they do become pregnant, I put them on caprylic acid and friendly bacteria and try to maintain the patients. It has become quite obvious to me that candida overgrowth has an anti-estrogenic effect on the body. This is probably the mechanism for the infertility.

More than 10 years ago, James Brodsky, M.D., a Chevy Chase, Maryland internist interested in yeast-connected health problems, began to share information with me about several women he had seen who were able to conceive after treatment with nystatin and the sugar-free diet.

He's careful to note, "I'm not saying that yeast causes infertility. Instead, I feel that any illness may be a factor. Sick people, whether due to stress, or just not being well for any reason, may be unable to conceive."

Dr. Brodksy notes the successful outcomes of two patients who were not able to conceive.

One patient had been trying to conceive for seven months, and the other for 10 months. On diet and nystatin, one woman became pregnant in 30 days and the other in 60 days. About the same time, I learned of another woman (I'll call her Evelyn) with yeast-related infertility. Even though her case is one of the older ones, it may sound familiar to many of you:

Evelyn's Story

I was unusually healthy as an infant and young child. My mother tells me I was rarely sick and I wasn't troubled by allergies. During my teen years, I was bothered by menstrual cramps—nothing unusual. I married early and my son was born when I was 19. No significant problems on through my 20s except for moderate menstrual problems and occasional yeast infections.

Four years ago, I married again and beginning three years ago, my husband and I decided we would like to have a child. Although we used no contraceptives, nothing happened.

I think that some of my health problems began to develop about 2½ years ago. At that time, I began low-mileage running and over a period of months, I lost about 15 lbs. I also developed other symptoms, including fatigue and lower back pain. I visited my family doctor who said, "You seem to have a kidney infection," and he prescribed Bactrim (a sulfonamide antibacterial medication used for urinary tract infections) for 10 days. Now as I think about it, I was premenstrual at the time, and the disappearance of my symptoms may have been related to my cycle rather than to the supposed urinary infection.

During the next six months, while attempting conception, I developed increasingly severe dysmenorrhea, continuing lower back pain and general fatigue. So I sought further help from a gynecologist who specialized in studying patients with infertility. About the same time, I began to have more acute pain on my right side with a lot of pelvic and abdominal bloating.

A laparoscopy revealed "mild/moderate endometriosis with a few adhesions" and chromotubation (this is an older method no longer in use)—which was attempted three times—showed that my right tube was closed.

I was placed on Danazol (a drug then used for endometriosis), and surgery was recommended. So I decided to seek a second opinion in regard to surgery and the quality of my husband's semen.

The following month, I began to have chronic uterine cramping, severe bloating, lower back pain, mood swings, nausea, occasional very painful right lower quadrant pain, insomnia, taut, dry skin (my skin was usually oily) and dizzy spells. I had so much pain that it scared me.

So I went back for further studies, and a hysterosalpingogram revealed that both of my tubes were open. Yet my symptoms continued. I was sick and emotionally and physically exhausted so my husband and I took a wonderful, but exhausting, two-week vacation.

Then I went to a university center and went through the infertility clinic. The head gynecologist thought I was "hyperestrogenic," which he said was not uncommon following a rigorous trip. Various studies were negative, including a test following intercourse.

When my symptoms continued, my doctor decided I was "severely depressed" and that she doubted that there was any organic basis for my symptoms. So she suggested I see a psychiatrist. I took her advice, yet, at the end of his evaluation, he said, "You aren't severely depressed and I don't think you need psychotherapy."

I was then seen by another infertility specialist at the same university center. He insisted that I be "scoped" again. This time they found "moderate endometriosis, but no adhesions." He recommended six months of Danazol. I developed a urinary tract infection following the laparoscopy, and I was treated with Bactrim and then ampicillin (an antibiotic).

I felt horrible for several weeks following this course of antibiotics. I attributed my continuing pain to endometriosis. Yet my doctor couldn't understand why Danazol hadn't eased the pain after I had taken it for over a month. (I think by this time he, too, decided I was an overanxious, depressed hypochondriac.)

Finally my pain abated and I began treating myself with vitamins and "friendly bacteria." I continued to have symptoms, including terrible headaches, bloating, junky discharge and vaginal itching. I was given Vibramycin, although my cultures for chlamydia and herpes were both negative.

Next I went back to the second university center for further studies and again received the diagnosis of mild endometriosis. They also said it was possible that I was hypersensitive to ovulation.

About that time I read *The Yeast Connection* and took the yeast questionnaire (see page XX). My score was 181, which indicated the candida almost certainly was playing a role in causing my health problems. Among the items in my history that added significantly to my score were repeated courses of antibiotics, birth control pills, worsened symptoms on damp days and sugar craving. I also experienced fatigue, the feeling of being drained, depression, abdominal pain, and PMS.

Based on this history, a physician consultant interested in yeast-related health problems said, "Your immune system and endocrine problems may be related to *Candida albicans*. And I think a therapeutic trial of nystatin, diet and nutritional supplements may help you."

In discussing my problem with me, he said, "Although candida is not the cause of your reproductive problems, it may be playing a significant role. And many women with hormonal dysfunction and reproductive organ symptoms will improve significantly on a simple, but comprehensive anti-yeast program extending over a period of six months to two years."

So I began on the nystatin. During the first two weeks, I showed a little lessening of my symptoms. Not a lot, but I was encouraged. After I'd been on the program four weeks, I was literally "a new woman." My depression, fatigue, bloating, mood swings, urinary symptoms and vaginal burning had all but vanished.

I talked to my doctor, who said, "Keep on with your nystatin. You can even cheat on your diet occasionally and see if it makes any difference." And he said that because of my history of vaginitis, that a vaginal anti-yeast suppository would be advisable.

Then the exciting news. Because I missed my period, I went back to my gynecologist who, following the examination, said, "Evelyn, you're pregnant!" Then I began to wonder if it was safe to continue the nystatin.

One of my gynecologists said, "Stop it. You could develop complications." Yet, my own gynecologist said to continue. I also called Dr. Orian Truss, who said that in his experience nystatin during pregnancy was safe.

My pregnancy proceeded uneventfully, and nine months and three weeks after starting on nystatin, my husband and I became the proud parents of a beautiful, healthy 8 lb. 9 oz. son. And in the birth announcement I sent to the open-minded doctor who advised my gynecologist to put me on nystatin, I said, "Thanks seems like too little to say for your part in helping us have this baby—we felt he should be named for you!"

He is a beautiful, healthy boy, and I am feeling good physically. I still have trouble believing my good fortune.

MY COMMENTS

Yeast overgrowth is certainly not "the cause" of infertility. Yet, for a couple struggling in vain to solve the problem, anti-yeast therapy is a safe option worth considering. Such therapy should be comprehensive and should feature probiotics, oral nystatin and a special, sugar-free diet. I would especially recommend this type of treatment program if either or both partners have a history that suggests yeast overgrowth may be present.

CHAPTER 21

Sexual Dysfunction

It's certainly not surprising that a woman with the multiple symptoms of candida yeast overgrowth would have little energy for or interest in sex.

Even if the inclination is there, frequently the pain from conditions like endometriosis, vaginitis, vulvodynia or interstitial cystitis may make even the thought, much less the act, of sexual intercourse unbearable.

There is little medical literature on sexual dysfunction and candida, but there is an interesting Australian study reported in the Australia-New Zealand *Journal of Obstetrics and Gynecology* in 1999.[1] It notes that candida was present in 21% of women participating in the study on vulvar vestibulitis syndrome (VVS), an inflammation of the area around the vagina, frequently present in women with vulvodynia and often a cause of painful intercourse (medically called dyspareunia). When those women received long-term antifungal therapy, 71% were cured, the article's abstract says.

Despite the dearth of medical literature on this subject, it is no secret that women with candida yeast overgrowth have sexual problems.

Of the thousands of women who have contacted me about their candida conditions, a large number reported sexual problems beyond the physical pain of sexual intercourse, including loss of interest and drive and inability to have an orgasm.

Complaints have been multiple and varied and usually include fatigue, headache, depression, digestive problems and PMS.

Many of my colleagues have confirmed that sexual dysfunction often occurs in patients with yeast-related health problems.

I'll share a few of these stories in hopes you may find some consolation in the idea that a complete anti-yeast regimen can relieve these problems.

A TYPICAL LETTER

In 1991, I received a letter from Loretta (not her real name), a public health nurse who was writing for information to share with patients who had yeast-related problems. In her letter, Loretta shared her own story of her recovery on diet and antifungal therapy:

Over the past four or five years, I began developing severe PMS. A week before my menses, I became irritable, with little things upsetting me that normally would not. I was also depressed and had no sexual desire for my husband.

After my menses ended, the irritability, breast pain and water retention resolved. The depression, difficulty thinking clearly, decreased libido and vaginal candidiasis remained . . .

My decreased libido was beginning to affect my marriage. I was just not interested in sex. I had no desire for it, which really caused problems for my husband. He felt I didn't find him pleasing to me, which caused him to be depressed. The old saying goes, "When Mama's not happy, ain't nobody happy." This was so true at our house.

Fortunately for me, I was introduced to The Yeast Connection. *I've been following the treatment program for almost a year now. With the diet modification, addition of vitamin supplements and nystatin, almost all of my symptoms cleared, except the vaginal candidiasis. Then I was able to obtain Diflucan, which completed the cure. I took a 200 mg. tablet for two days and have not had a recurrence. Finally, the cycle was broken.*

Now at my house, everybody's happy, because Mama's happy. Our lives have most definitely been changed for the better.

A REPORT ON ONE OF MY PATIENTS

In March 1993, I saw 33-year-old Marjorie in consultation. Her main complaints included "always tired and drained and loss of sex drive (four years duration); yeast infections/vaginitis (two years duration)." Here are excerpts from a letter Marjorie sent me prior to our first consultation:

All of my life I've been troubled by many nose and throat infections and have taken many antibiotic drugs. In 1988, my son was born, after a difficult labor. After that, it seemed my health went downhill. Although I went back to work, I had to quit after three months. I just couldn't do it. I didn't have the energy. I also had a marked loss of sex drive.

Then came the yeast infections, vaginal dryness, urinary frequency, PMS, cramps, depression, sleep problems, abdominal pain and very poor memory. I saw a number of doctors who gave me different medications, including antidepressants. One doctor said, "I can't find anything physically wrong." My tests were all negative.

I know that I look terrible, feel terrible and act terrible and have absolutely no sex drive. If I didn't have such a loving husband I don't know what I would do. Please, I need help for me and my son.

Because of the severity of Marjorie's symptoms, I recommended Diflucan, a sugar-free diet and nutritional supplements. During the succeeding months, Marjorie improved and on April 14th, just six weeks later, she reported:

The vaginitis and bladder pain are gone. Yet, I still have some rectal itching. I still have good days and bad days, especially when I overdo. Sex is better. I'm not as irritable or moody. I wake up rested most of the time and exercise daily. My memory is a lot better. I still have a way to go and I miss my coffee and sandwiches.

Then on August 23, 1993, less than six months after her first visit, Marjorie said:

I'm feeling much better now. I'm just a new person altogether! My patience, memory, energy level, depression, sex drive and sleeping have improved 100% since the first time I saw you . . . But I really have to stay on my diet. I also have to exercise, which I do three or four times a week. It really helps. I can actually do a day's work without "feeling completely worn out." My PMS is much better, but it's still there.

. . . In spite of these problems, my life and my marriage are

much better. Thank God for the help I've received. I know I still
have a ways to go before I'm well, but it's closer than before.

On September 29, 1993, Marjorie came in for a brief visit and she
said:

I'm feeling great, absolutely great! As I told you last month, I
really am a new person. Something I haven't told you before. I
really felt so bad that life really wasn't worth living, and suicide
was an option I thought about from time to time.

Another factor that added to my load was the problem I had
coping with my hyperactive, irritable, inattentive 4-year-old son.
Nothing I had done for him helped . . . until you put him on the
diet, Diflucan and nutritional supplements. And the changes in his
behavior and ability to learn have truly been remarkable. Thank
you, thank you, thank you!

The Other Side of the Coin

Sometimes yeast-related problems manifest themselves in what
seems like exactly the opposite way.

The story of Patty's (not her real name) sexual nightmare actually
sparked other letters from women who had suffered similar symptoms.
It began when she was 35. Her symptoms included:

. . . homosexual feelings or intense heterosexual feelings that would
persist for hours, even days . . .This constant state of sexual arousal
became an obsession with me. I could not concentrate on anything else.

Keep in mind that during this time most of my severe inner
conflicts were religious ones. I was raised in a Catholic church with
very strict moral standards. I was a virgin when I married at the
age of 22, so I found the anguish of these uncontrollable urges and
thoughts to be emotionally destructive.

During times of anger, my body and my emotions would
respond inappropriately with intense feeling of sexual arousal. Bear
in mind that this sensation was insatiable. Even "satisfactory"
regular intercourse did nothing to control or reduce this constant
sexual frustration . . .

*I went to dozens of doctors, including psychiatrists, all to no
avail. Finally, I found a doctor to help me with a yeast problem,
and my terrible nightmares began to improve.*

The next time I heard from Patty, she reported she was 95% better
and no longer suffered from "horrifying sexual abnormalities" and required "no more psychotherapy or mood-changing drugs."

I've stayed in touch with Patty over the years. She now works as an
executive of a nonprofit health organization. In a recent letter, she said:

*I'm happy to report that I'm healthy and happy, and my sexual
nightmare is now only an unpleasant memory.*

During the past several years, I've received hundreds of letters and
calls from women describing their sexual dysfunction, and among
those letters were some who said that reading Patty's story in the first
edition of *The Yeast Connection and the Woman* made them feel better
because they knew they were not alone.

MY COMMENTS

Based on the observations of both professionals and nonprofessionals, sexual dysfunction (including decreased libido and orgasmic dysfunction) occurs frequently with yeast-related problems. And in many
women, a sugar-free special diet and prescription and/or nonprescription antifungal medication help. Yet, as I have expressed elsewhere in
this book, such treatment will not be a quick fix and should be combined with other therapies that may include nutritional supplements,
control of environmental pollutants and possibly thyroid and other endocrine therapies. Psychological counseling may also be of help in addressing this condition.

REFERENCE

1. Pagano R., "Vulvar vestibulitis syndrome: an often unrecognized cause of
dyspaeunia." Australia-New Zealand *Journal of Obstetrics and Gynecology,* 1999.

CHAPTER 22

Autoimmune Disorders

"**A**uto" is the Greek word for self. The immune system is a complicated network that works to defend your body and eliminate infections, viruses and invasions of all types of microbes. It takes care of challenges of all sorts, ranging from colds and flu to wounds, and even cancer.

Autoimmune diseases occur when the immune system becomes confused and attacks the body, targeting its own cells, tissues and organs. When immune cells collect at a target site, inflammation occurs.

There are dozens of autoimmune diseases, each affecting the body in a different way. An autoimmune reaction targeting the brain and the nervous system manifests as multiple sclerosis (MS), and when it takes place in the gut, it's called Crohn's disease. The immune system's attack on joints manifests as rheumatoid arthritis. A systemic autoimmune reaction takes place in lupus (systemic lupus erythematosus), affecting anything from skin and joints to kidneys and lungs to the brain. Type 1 diabetes (once known as juvenile onset) is the result of the immune system's attack on insulin-producing cells in the pancreas.

In this section, we'll talk about four autoimmune diseases that most commonly affect women.

As a group, autoimmune diseases afflict millions of Americans, 75% of them women. Women of working age and in their childbearing years are particularly prone to these diseases.

The most common autoimmune diseases from which patients with yeast overgrowth suffer are multiple sclerosis and myasthenia gravis.

Again, I'm not saying yeast overgrowth causes autoimmune diseases, but I am suggesting there might be a link somewhere.

MULTIPLE SCLEROSIS

The Multiple Sclerosis Society calls MS "a chronic unpredictable neurological disease" that affects an estimated 400,000 Americans. Women are 2.6 times more likely than men to have MS, but the disease progresses more quickly in men. It's the fourth-leading cause of disability in American women. Most victims are in their 40s and 50s.

Alarmingly, the number of women with MS has dramatically increased in the past 20 years, jumping 50% between the 1980s and the 1990s without explanation, according to the American Autoimmune Related Diseases Association.

Medical science suggests multiple sclerosis is caused by the immune system turning against the tissues of the brain and the spinal cord. When the immune system kicks in, it strips the neurons of their protective insulation, making simple actions such as walking and talking more difficult, usually resulting in fatigue, tremor and paralysis. This normally takes place when the patient is between the ages of 20 and 40. Science has not yet found a cure for MS, but there are treatments.

Symptoms

These symptoms may be permanent or they may come and go:

- extreme fatigue
- blurred vision
- loss of balance
- poor coordination
- slurred speech
- tremors
- numbness
- problems with memory and concentration
- paralysis (in severe cases)
- blindness

Treatment

Immune-system-suppressing drugs can reduce the severity of the symptoms, but they don't cure the disease.

Stress can worsen MS symptoms, and a general wellness program, including exercise, a healthy diet and stress management are part of the standard treatment for MS, along with prescription medications that can ease symptoms.

MYASTHENIA GRAVIS

Myasthenia gravis is similar to MS, but they are on the surface only. It is a deficiency of neuromuscular transmission affecting about 36,000 people in the United States. Most of them are women between the ages of 20 and 30, and the disease gradually affects more men after age 50.

Autoimmune attacks on the nerve-muscle junction are believed to cause myasthenia gravis (sometimes known as Aristotle Onassis' disease).

Symptoms

- muscle weakness, but not usually fatigue
- eye muscle disturbances
- drooping eyelids
- difficulty chewing, swallowing or talking
- tumors on the thymus gland
- blurred vision
- weakness in arms or legs
- unstable or waddling gait

Treatment

There is widespread disagreement in the medical profession about the treatment of myasthenia gravis.

Spontaneous improvement and even remission are possible, especially in the early stages of the disease.

Prescription drugs such as cholinesterase inhibitors (Mestinon and Prostigmin) are helpful for some patients, corticosteroids and immunosuppressant drugs are helpful to others, but all three can have serious side effects. Other possible treatments include intravenous immune globulin treatments, plasma exchange and surgical removal of the thymus gland.

RHEUMATOID ARTHRITIS

Rheumatoid arthritis is an autoimmune disease that involves inflammation in the lining of the joints. The inflamed joint lining, called the synovium, can invade and damage bone and cartilage, resulting in pain, stiffness and swelling.

Rheumatoid arthritis affects about 2.1 million Americans, 1.5 million of them women. Interestingly, recent studies from Harvard's Brigham and Women's Hospital in Boston shows that women with rheumatoid arthritis have about twice the risk of a heart attack as women without the disease. Researchers theorize the link is because of the inflammation that is present with arthritis and is thought to contribute to the fatty build up in blood vessels, one of the known causes of heart attacks.

Symptoms

- inflammation of joints
- swelling
- difficulty in movement
- joint pain
- loss of appetite
- fever
- loss of energy
- anemia
- sometimes rheumatoid nodules (lumps of tissue under the skin)

Treatment

RA can be diagnosed with a blood test for an antibody found in approximately 80% of people afflicted with the disease.

Drug treatments include those that help relieve symptoms, such as NSAIDs (non-steroidal anti-inflammatory drugs), aspirin, prescription pain relievers and glucocorticoids to help reduce joint pain, stiffness and swelling.

Sometimes patients are given corticosteroid medications such as cortisone or prednisone for relief during severe flare-ups.

Patients with rheumatoid arthritis who have failed to respond adequately to NSAIDs, may want to consider a different class of medication called disease-modifying drugs or DMARDS. This category of drugs includes hydroxychloroquine, gold, sulfasalazine, minocycline, methotrexate, penicillamine and leflunomide.

PSORIASIS

Psoriasis is a painful skin disease that got its name from the Greek word for "itch." The skin becomes inflamed, with red, thickened areas with silvery scales, most often on the scalp, elbows, knees and lower back. It affects 4.5 million Americans.

It's believed to be caused by an autoimmune response that sends faulty signals that speed up the growth of skin cells, resulting in non-contagious patches of red skin covered by a flaky white buildup. Certain types of psoriasis may manifest as a pimply surface with a red and burned-looking appearance.

About 30% of psoriasis sufferers will develop a related form of arthritis, called psoriatic arthritis.

While psoriasis appears to occur equally in men and women, women with the disease have additional concerns because some of the medications used to treat it can cause birth defects, which means they can't be used by women of childbearing age.

Symptoms
- flaking skin
- redness
- itching
- swelling
- can cover anywhere from 2% to 100% of the body.

Psoriasis is aggravated by emotional stress, injury to the skin, some types of infections and reactions to certain drugs.

Treatment
Treatment usually takes place in steps, depending on the extent and seriousness of the disease. It's sometimes called a "1-2-3" approach. In Step 1, medicine is applied to the skin. Step 2 uses light therapy and in Step 3, patients are given oral medications.

Two drugs for severe psoriasis, etretinate (Tegison) and isotretinoin (Accutane), should never be taken by women who plan to have children. Another treatment, PUVA, combines a type of drug called a psoralen with exposure to ultraviolet A (UVA) light, and is also considered inadvisable for pregnant women.

Exposure to sunlight, soothing baths and moisturizers is helpful for some patients.

THE YEAST CONNECTION AND AUTOIMMUNE DISORDERS

There are many theories on the origins of autoimmune disorders, but anecdotally, I have discovered that, in some cases, antifungal medications and sugar-free diets for patients with some of these symptoms provide relief as long as the treatment continues. The symptoms usually return as soon as the treatment is discontinued.

It has been my experience that the anti-yeast treatment helps many patients, especially those with symptoms of MS.

In discussing multiple sclerosis and other autoimmune disorders in *The Yeast Connection,* I said, "Candida isn't THE cause of these often devastating disorders—but, there's growing evidence based on exciting clinical experiences of many physicians that there is a yeast connection."

Dr. Truss had a similar experience. In his observations on candida-related health problems, Dr. Truss described the response of a number of his patients with severe autoimmune diseases to nystatin and a low-carbohydrate diet. Included were brief descriptions of several of his patients with multiple sclerosis.

One of these patients was a 30-year-old woman who showed many of the symptoms and signs of MS, including numbness, tingling, reflex changes, visual defects and a slight elevation of her spinal fluid protein. In addition, she gave a history of receiving many antibiotics, digestive problems, vaginal symptoms and other health problems. She was placed on a diet and received nystatin therapy, and after two years, her neurological examination was "entirely normal."

When the therapy was discontinued, a number of her symptoms began to return. She was again placed on nystatin and in a subsequent report, Dr. Truss stated:

"She is entirely well now, seven years after nystatin was begun."

OTHER PHYSICIANS' VIEWPOINTS

Over the years, I've corresponded with Cincinnati neurologist R. Scott Heath, M.D.

Dr. Heath has subscribed to the theory of a candida/autoimmune disease connection:

Sometimes I've felt sort of like a voice in the darkness . . . just as you have felt. Yet, I have seen a number of people who have unusual symptoms, including fleeting numbness, transient speech problems and vision complaints. In such patients, I often can't find anything wrong neurologically. Obviously my concern is simply this: Do they have MS or don't they?

I've been impressed over the years that people with MS who do not have spots on their brain generally do well on the anti-fungal treatment. Also, when people DO have spots and we make a diagnosis of MS, they do better, too. They have fewer exacerbations . . .

Today if a person is seriously ill with an exacerbation, I'll put them on Diflucan. Then once they're off the prednisone or ACTH, I'll switch them back over to nystatin. The problem is it's kind of hard to judge exactly what I'm doing . . . I'm sort of a lone wolf in my group of 15 neurologists. None of my colleagues give it a whole lot of credence.

Dr. Heath also suggests people with autoimmune diseases may have exceptionally sensitive immune systems:

There's something peculiar about MS patients and/or some of these CFS/CFIDS patients, as well as people with other auto-immune diseases. In such patients, whatever the reason, the immune system is very sensitive. I don't yet think we have the answers. I do believe that the yeast factor is at least part of the answer . . .

Other physicians, like Zoltan P. Rona, who practices nutritional medicine in Toronto and is past president of the Canadian Holistic

Medical Association offers a theory about what might be taking place. He notes that Dr. Truss thinks much of the harm done by candida comes from its waste product, acetaldehyde:

". . . Few chemicals can create so much havoc in the body as acetaldehyde can . . . *acetaldehyde is a fungal waste product . . .*"

In addition, there is the theory espoused by some scientists that *Candida albicans* floods the system with an accumulation of its waste product—the toxic acetaldehydes—which then poison the tissues, accumulating in the brain, spinal cord, joints, muscles and tissues.

The renowned diet doctor, the late Robert Atkins, M.D., has also made a connection. He recommends vitamin B5 (pantothenic acid) to aid in the growth of friendly bacteria like lactobacillus bulgaricus and bifidobacterium. Dr. Atkins said, "In people with candidiasis, (vitamin B5 in the form of pantethine) fights off a toxic byproduct called acetaldehyde . . . which, without toxic consequences can reduce cholesterol, counteract oxidation, stimulate the growth of friendly bacteria and fight allergies, inflammation, autoimmune disruptions and alcoholism."

In an article in the December 2001 issue of *Let's Live,* "Nutritional Rx for Multiple Sclerosis," nutritional expert Jack Challem said, "Conventional medicine can't do a lot for people with MS . . . The solution may be to become your own healer."

He suggests that a simple, natural diet and specific supplements to help many people with MS, and he recommended a high-protein, low-carbohydrate "paleolithic" diet—Omega-3 fatty acids, antioxidants and vitamin B^{12}. And in discussing food allergies and nutrient deficiencies he said:

The body's immune system generally ignores food proteins but reacts to foreign proteins (e.g., viruses and bacteria). Occasionally, in a process called molecular mimicry, harmless proteins are so similar to harmful ones that the body reacts to them. This accounts for many food allergies, and immune response to the casein in milk and gluten in wheat and other grains have been well documented in other diseases. A similar process occurs in some cases of MS.

Research has pointed to a lack of vitamin D, omega-3 fatty acids, and other nutrients in those with MS.

Joyce's Story*

In August 1992, this 35-year-old woman called me and said,

I've had severe myasthenia gravis for over 20 years and I have been troubled by many other symptoms. I recently obtained a copy of your book, The Yeast Connection, *and began to change my diet. I'm already improving! I'm off all MG medication and gaining back my strength.*

Joyce also told me she'd taken large amounts of antibiotics, and that she had been troubled by recurrent vaginal yeast infections in childhood and early adolescence. At age 13, she began to show symptoms that led to the diagnosis of MG. Her symptoms at that time included: completely nasal voice—air came out of her nose instead of her mouth when she spoke; she was unable to swallow and had to wash down food with liquid. Joyce had no control of the muscles in her face. She was unable to smile, and her eyes remained open when sleeping—she was uncoordinated and troubled by double vision, migraine headaches and stomachaches.

During the past 10 years, I've exchanged many letters and e-mails with Joyce, and her story is truly remarkable. She has regained her health and her life.

Like many people with yeast-connected problems, she's had her ups and downs and had to do things differently. During 2001 and 2002, she found that taking Diflucan helped her more than nystatin. She still follows her diet, which features a variety of foods, including vegetables, chicken, fish and seeds.

MY COMMENTS

Are all autoimmune diseases yeast-related? I don't really think so. Yet, the observations of Dr. Truss and those of other physicians suggest

---

*You can read more about Joyce's story in *Yeast Connection Success Stories.*

that many of them are, and that a comprehensive treatment program that includes an appropriate diet, probiotics and antifungal medication can help many people.

It is important to review Chapter 19 when considering auto-immune disease. Reducing exposure to chemicals is the recommenda-tion of neurosurgeon Dr. Russell Blaylock, who reminds us that the body will not attack a healthy self, but only a toxic or unhealthy self.

Joyce's story illustrates many points relevant to the person with a chronic health disorder, including those which are yeast-related. Here are a few of her points that I would like to emphasize:

- Although many chronic health disorders are yeast-related, antifungal medications alone do not provide a "quick fix" or a long-term answer.
- An appropriate diet is essential for any woman who wishes to re-gain her life and health.
- Faith in God, courage, persistence and determination are key in-gredients in any treatment program.

RESOURCES

- American Autoimmune Related Diseases Association
  http://www.aarda.org
  (800) 598-4668

- Arthritis Foundation
  http://www.arthritis.org
  (800) 283-7800

- Crohn's and Colitis Foundation of America
  http://www.ccfa.org
  (800) 932-2423

- Lupus Foundation of America
  http://www.lupus.org
  (800) 558-0121

- Mysathenia Gravis Foundation of America
  www.myasthenia.org
  (800) 541-5454

- National Multiple Sclerosis Society
  www.nationalmssociety.org
  (800) FIGHTMS

- National Psoriasis Foundation
  http://www.psoriasis.org
  (800) 723-9166

Shomon, Mary. *Living with Autoimmune Disease,* HarperResource, 2002.

PART 4

Yeast-Related Disorders That Affect Both Sexes Equally

CHAPTER 23

Food Allergies and Sensitivities

know I've mentioned this story before, but it bears repeating here with a few more details.

When I completed my internship and residency training and returned to Tennessee to open my first medical office, I knew nothing about hidden food allergies.

I continued to be a food allergy ignoramus until the mid-1950s when Aileen, the mother of a 12-year-old boy, opened my eyes. Here's what happened: Her son, Tom, was so tired she could hardly get him out of bed for school each morning. Tom also complained of headaches, belly aches and muscle pains, and often was so irritable, Aileen said, "You can hardly stay in the house with him."

Because Tom had experienced problems with milk in infancy and had been drinking a lot of milk for several months before these complaints developed, Aileen took him off milk for a week. At the end of the period, she said, "Tom is like a different child. He bounced out of bed this morning whistling. No headaches, muscle aches or belly aches."

> I was astounded because I had learned something that I hadn't known before . . . intolerances, allergies or sensitivities to common foods could provoke systemic and nervous complaints.

A short time later, I read several articles in the medical literature in which physicians reported that many of their patients improved—often dramatically—when they avoided wheat, corn, milk, eggs and other foods.

165

I collected and summarized my findings on 50 of these patients and published them in a major pediatric journal more than 30 years ago. Here's how the diagnosis of food allergies/sensitivities was made: Symptoms and signs were relieved by eliminating suspected foods from the diet for 5 to 12 days and then reproduced by giving the food back to the child, and noting the child's reactions.

Over the years, I discovered that carefully designed and properly executed elimination diets helped many of my patients. I've also found that such sensitivities can affect almost every part of the body, and they occur commonly in women with yeast-related disorders.

Unusual reactions to substances in a person's diet or environment have been recognized for thousands of years. Yet, it wasn't until 1906 that the term "allergy" was coined by the Austrian pediatrician Clemens von Pirquet.[1] He put together two Greek words, *allos*—meaning "other" and *ergon*—meaning "action." To von Pirquet, allergy meant altered reactivity.

Some doctors feel the term "allergy" should be limited to those conditions in which an immunological reaction can be demonstrated using allergy skin tests or more sophisticated laboratory tests. But other knowledgeable physicians feel that the allergic and hypersensitivity diseases are much broader. For example, Elmer Cranton, M.D., a Yelm, Washington, past president of the American College of Advancement in Medicine, said:

> The assumption is often made that we fully understand the immune system, which is not true. Although these published scientific discoveries add to our knowledge, it's quite possible we still only know a small fraction of the whole story.[2]

SYMPTOMS

Most of us are familiar with the itching, sneezing, stuffy-headed response some people have to irritants of all kinds, ranging from pollens to mold spores to animal dander to foods and medications. Allergies are often characterized as a hypersensitivity to certain substances. Here are some symptoms:

- watery, itchy eyes
- sneezing
- runny nose, stuffy head
- hypersensitivity pneumonitis, usually caused by exposure to organic dusts like bird droppings, feathers and contaminated grain
- eczema
- dermatitis
- hives
- itching
- anaphylactic shock, the most severe and life-threatening, causes swelling of body tissues, including the throat and a sudden drop in blood pressure, that often occurs in cases of extreme sensitivity to penicillin, stinging insects, shellfish or nuts.

TREATMENT

Treatment for allergies starts with avoidance of the allergen.

Over-the-counter or prescription antihistamines relieve and sometimes prevent the symptoms of allergic rhinitis (hay fever) and some other allergies. Decongestants can help shrink the blood vessels and relieve nasal congestion.

WHY YOU NEED TO KNOW ABOUT FOOD ALLERGIES/SENSITIVITIES

I've said this before and it bears repeating here: Every person with a yeast-related problem has an overgrowth of *Candida albicans* in the digestive tract. This creates a disturbance in the normal balance of good bacteria, which, in turn, leads to a weakness of the membrane lining the intestinal tract and what is commonly called a "leaky gut." As a result, antigens (or toxic substances) in food are absorbed, and this plays a part in making you sick.

More than a decade ago, J. O. Hunter, a fellow of England's Royal College of Physicians, in a discussion of irritable bowel syndrome, suggested that patients with food intolerances have an abnormal gut flora, even though pathogens may not be present. In a discussion of food intolerance in the British journal, *The Lancet,* he said that food allergies may actually be caused by fermentation in the colon and by correcting imbalances in the gastrointestinal flora, those intolerances may be resolved.[3]

Still other investigators, including Leo Galland, M.D., and W. Allen Walker, professor of pediatrics at Harvard Medical School, add that molecules that penetrate the intestinal walls (we've called it "leaky gut") may not have "nutritional importance," but they may certainly have "immunologic importance."[4,5]

Types of Allergies

When you develop an allergy to something you breathe, such as grass or ragweed pollens, animal danders or house dust mites, the cause of your symptoms can be suspected from your history and identified by the use of the simple allergy prick test. In carrying out such a test, a physician pricks your skin through a small amount of an allergy extract.

Yet, most people with food allergies and sensitivities do not show a positive prick test, and these allergies are often called "hidden," "masked," "variable" or "delayed-onset" food allergies. Allergies or sensitivities of this type are often caused by foods you eat every day. You may be surprised to learn that you're apt to be sensitive to some of your favorite foods, especially wheat, corn, milk, yeast, chocolate, citrus and coffee.

Moreover, you may be "addicted" to foods that are making you tired or cause headaches, muscle aches or nasal congestion. Like the cigarette or narcotic addict, you may feel better for an hour or so if you've eaten some of the foods to which you're allergic.

Eight foods cause 90% of all allergic food reactions, according to the Asthma and Allergy Foundation of America: milk, soy, eggs, wheat, peanuts, tree nuts, fish and shellfish.

The Foundation says, "People with food allergies have supersensitive immune systems that react to harmless substances found in food and drink."

Sounds familiar! Re-read the last chapter on autoimmune disorders and consider the theory that people with these diseases have hypersensitive immune systems. We begin to pick up the thread of the yeast connection this way.

OBSERVATIONS OF OTHER PROFESSIONALS

In the his award-winning presentation delivered before the Section on Allergy of the American Academy of Pediatrics, William C. Deamer, M.D., of San Francisco, commented on the deceptive nature of food allergies and how difficult they are to diagnose and treat. He believed that skin tests to determine food allergies are unreliable and advocated elimination diets when he was looking for causes of fatigue, irritability, headaches, abdominal pain, musculoskeletal discomfort, asthma and other unexplained symptoms in his patients. And he said:

> There can be no doubt . . . of the role specific foods may play in causing these symptoms . . .[6]

Ted Kniker, M.D., of San Antonio, Texas, said in an award-winning article:

> There are countless millions of individuals who have unrecognized adverse reactions to various antigens, foods, chemicals and environmental or occupational triggers . . . The acquired disease may be limited to body surfaces, or may involve a puzzling array of organ systems causing the patient to visit a number of different specialists who are unsuccessful in recognizing that an allergic or adverse reaction is going on.[7]

SUPPORT BY PRACTICING PHYSICIANS

Hundreds of practicing physicians have found that they can help many of their chronically ill patients by identifying and eliminating foods that play a major role in causing their symptoms. One such physician, my good friend Dr. Elmer Cranton, has for more than a decade used elimination/challenge diets in treating his patients with yeast-related health problems. In a recent letter, he said to me:

> I routinely prescribe elimination/challenge diets in all of my patients with yeast-related problems. By identifying trouble-making foods and removing them from the diet, I have decreased the patient's allergic load. Then the antifungal diet and other parts of my treatment program are much more apt to be successful.[8]

Many other physicians who are members of the American Academy of Environmental Medicine, the American Academy of Otolaryngic Allergy, the Pan American Allergy Society and the American Holistic Medical Association, use elimination/challenge diets in helping their chronically ill patients.

So do a small but growing number of physicians who are members or fellows of the American Academy of Allergy and Immunology, the American College of Allergy and Immunology and the American Academy of Family Physicians.

One member of the latter organization, Dr. Harold Hedges, of Little Rock, Arkansas has for more than a decade, emphasized the importance of delayed-onset hidden food sensitivities. In a comprehensive article in the November 1992, *American Family Physician,* he said:

> The connection between food and a number of medical illnesses is clearly established. Adverse reactions to food may also be responsible for frequently recurring or chronic symptoms in patients whose medical examinations and tests are normal. In these and other patients, a well-planned and well-executed elimination diet can be the key to diagnosis. Specifically, use of an elimination diet might be considered when no other cause can be found for the symptoms.[9]

SUPPORT IN THE PRESS

The frequency and importance of food allergies was emphasized by health columnist Jane Brody in *The New York Times* who said:

Millions of Americans say certain foods make them sick. Are doctors paying close enough attention?

She also discussed the controversy about food allergies and sensitivities among physicians and said:

> In researching this article, I initially believed that most of the claims attacking this food or that as the cause of everything from hair loss to athlete's foot were elaborate hokum. But after looking at the medical research and learning about various people's experiences, I now wonder whether the rigid thinking of some doctors is not ill advised. Indeed, in dismissing symptoms that don't involve the immune system, these doctors might be doing a disservice to the health and well being of millions of Americans.[10]

Jean Carper, whose books and columns have been read by millions of Americans, has discussed food allergies in her weekly column in *USA Weekend,* which I've read for the past several years. And in her book, *Food—Your Miracle Medicine,* she lists headaches, asthma, irritable bowel syndrome, chronic fatigue syndrome and depression as complaints that may be food-related. She also said that there's growing scientific recognition that these maladies are often food-related.[11]

HOW TO TRACK DOWN YOUR HIDDEN FOOD SENSITIVITIES

Over the years, as I worked with women with yeast-related problems, I learned that almost without exception, every woman with these problems was bothered by food sensitivities.

Here are some of the things you'll need to do to track down what may be aggravating your problems:

- Before beginning the diet, sit down with family members and get their cooperation.

- Carry out the diet at an appropriate time. (Don't try it when you're traveling or during a holiday.)
- Before beginning the diet, you'll need to keep a symptom diary and inventory for at least three days.
- Continue a diet diary while you remain on the diet.
- Continue the diet for five to 10 days or until you show convincing improvement in your symptoms.

See www.yeastconnection.com for a printable food record/symptom inventory to help you identify food sensitivities.

- To identify food troublemakers, return the eliminated foods to your diet, one food each day, and see if your symptoms return. You may notice symptoms within a few minutes, or they may not occur for several hours or until the next day.

On the initial diet you should avoid:

- milk
- Kool Aid, punch, etc.
- wheat
- eggs
- chocolate
- sugar in all forms
- corn
- food colors and dyes
- soft drinks
- processed and packaged foods

If you've been bothered by asthma or have experienced swelling or other serious reactions, get the help and consultation of your physician before carrying out the diet.

Different allergy troublemakers may bother you, and other factors may lower your resistance. I sometimes refer to them as "gremlins."

To regain your health you'll need to discover which "gremlins" are causing your problems and take steps to control them.

MY COMMENTS

Over the past 15 years, researchers have found that people who experience hidden or delayed-onset food allergies may often show abnormal tests. Although I've used these tests from time to time, I've relied more on the elimination/challenge diet, and I find the results are much more beneficial to my patients this way. However, if you just don't have the time to do the elimination/challenge diet, the allergy tests can be invaluable. See Chapter 6 for a list of labs that perform food allergy tests.

And, beginning in 2002, I now recommend electrodermal screening (also known as BioMeridian testing) as an additional aid to tracking down food sensitivities. (See Chapter 34.)

REFERENCES

1. Von Pirquet, C., "Allergie," *Munch Med. Wochenschr.*, 1906; 53:1457.

2. Cranton, E., Personal communication, July 1994.

3. Hunter, J. O., "Food Allergy—or Enterometabolic Disorder?," *The Lancet,* 1991; 338:495–496.

4. Galland, L., "The Effect of Microbes on Systemic Immunity," in Jenkins, R. and Mowbray, *Post-viral Fatigue Syndrome,* John Wiley and Sons, 1991.

5. Walker, W. A., in Brostoff and Challacombe, *Food Allergy and Intolerance,* London, Balliere Tindall, Philadelphia, W. B. Saunders, 1987; pp. 209–222.

6. Deamer, W. C., "Some impressions gained over a 37 year period," *Pediatrics,* 1971; 48:930.

7. Kniker, W. T., "Deciding the Future for the Practice of Allergy and Immunology," *Annals of Allergy,* 1985; 55:102.

8. Cranton, E., Personal communication, July 1994.

9. Hedges, H. H., "The Elimination Diet as a Diagnostic Tool," *American Family Physician* (Suppl) 1992; 46:77–84.

10. Brody, J., *The New York Times,* April 29, 1990.

11. Carper, J., *Food—Your Miracle Medicine,* HarperCollins, New York, 1993

CHAPTER 24

Asthma and Allergies

More than 50 million Americans suffer from asthma and allergies. Of these, 17 million have asthma, a potentially life-threatening disease that affects airflow when breathing muscles squeeze, swell or are blocked by excess mucus.

More than 70% of asthma sufferers also have allergies. The Centers for Disease Control and Prevention says that the prevalence of asthma has increased dramatically in recent years, up 75% in just 14 years, between 1980 and 1994. Experts theorize the increased amount of air pollution is responsible for at least part of this increase.

CDC statistics also show that 9.1% of women suffer from asthma, and 5.1% of men have the disease.

SYMPTOMS

In an acute episode, the following symptoms are most frequently present:

- wheezing
- coughing
- shortness of breath
- constriction of the chest muscles
- sputum production
- excess rapid breathing/gasping
- rapid heart rate
- exhaustion

Other symptoms that may be noticeable, both during an acute attack or at any time, include:

- chronic nighttime coughing
- chest pain
- chest tightness
- insomnia due to shortness of breath
- acid reflux disease
- sensitivity to medications, especially NSAIDS (such as ibupro-fen) and beta-blockers used to treat heart disease and migraines
- *intolerance to the smell of chemicals*

Asthma symptoms can be triggered by:

- allergens, including food allergies
- viral or sinus infections
- exercise
- emotional anxiety
- air pollutants such as tobacco smoke and airborne chemicals
- workplace allergens, dust, vapors or chemical fumes
- strong odors or sprays such as perfumes, household cleaners and chemicals
- changing weather conditions

TREATMENT

Standard treatment of asthma involves continuous use of medications to prevent airway inflammation. Among them are:

- corticosteroids, both inhaled and taken orally.
- anti-inflammatory medications, such as cromolyn or nedocromil, that prevent lung inflammation and stop inflammation if it occurs.
- rescue medication taken to stop an attack in progress, including beta-agonists and theophylline.
- anti-leukotrienes that fight the leukotrienes responsible for airway inflammation.

THE YEAST CONNECTION

Based on the research studies of Iwata and Witkin in the 1960s and 1970s, and the clinical observations of Dr. Truss, superficial yeast infections may adversely affect the immune system, which, in turn, may adversely affect many different parts of the body. Yet, I'd never thought about the yeast connection to asthma until a physician friend added a note to a Christmas card I received in the early 1990s that said, "One of my patients with intrinsic asthma showed a marvelous response to Nizoral."

My interest in the yeast connection to allergies increased when I read two 1994 abstracts published in the *Journal of Allergy and Clinical Immunology*.[1] One of these abstracts cited the observations of Belgium researchers who described the favorable response of some of their asthmatic patients to (ketoconazole) Nizoral. *They found that four out of five of their asthmatic patients treated with this antifungal drug improved, while four out of five of the placebo group did not improve.*

A second abstract, published by researchers at the University of Virginia, described a study of 10 patients with asthma and associated fungal infections of their feet. In a randomized study, they found that eight of the ten asthmatic patients were able to reduce their steroid dosage with no adverse reactions to fluconazole (Diflucan), in a dose of 100 mg. daily, were noted in any of the 10 patients, including "those maintained on fluconazole for up to two years."[2]

In a June 2002 phone visit with Dr. George Ward, the senior author of the 1999 Virginia study, he told me that fluconazole may help an occasional patient with asthma who does not have a skin fungal infection. He also said, "When it works, it works very well . . . It never caused problems in my patients, and I do not hesitate to use it . . .

> There are patients with asthma who have a fungus problem, and it hasn't been realized enough for most physicians to be comfortable with it . . . I think that physicians will come to realize that it can be a therapeutic agent when they've looked at everything else. I certainly agree that fungi have something to do with asthma.[3]

Further support for the fungal connection to asthma was presented by Talal Nsouli, M.D., a Washington, D.C. allergist/immunologist at the 1999 annual conference of the American College of Allergy, Asthma and Immunology. Dr. Nsouli described the response of a 36-year-old man with "unstable, severe, recalcitrant, corticosteroid-dependent asthma," who also had multiple brown skin lesions (over an inch in length and width) on his back and chest. RAST testing was positive and skin testing was strongly positive to *Candida albicans*. Here's an excerpt of their summary:

> We started the patient on a trial of fluconazole for three weeks and clotrimazole topical cream for eight weeks. This resulted in an unanticipated dramatic and striking improvement of his asthmatic condition, as we were able to gradually taper him off systemic corticosteroids without deterioration of pulmonary function. His skin lesions cleared. He has not had any emergency room visit or hospitalization for the last two years. We conclude that dermatomycosis (a fungal infection of the skin) should be considered as a possible allergenic trigger in patients' unexplained recalcitrant intrinsic asthma.[4]

EARLIER REPORTS ON THE CANDIDA-ASTHMA CONNECTION

In reviewing the medical literature I found that the relationship of *Candida albicans* to asthma was noted by Itkin and Dennis (The National Jewish Hospital in Denver) almost 30 years ago. In the introduction of their paper, these investigators said:

> It is the scientific aim of the allergist to reduce the number of patients who must be classified as suffering from "asthma of unknown origin." It is the purpose of this paper to describe three years of experience with provocative bronchial challenge by inhalation as a tool in establishing *Candida albicans* as a significant allergen . . . in patients suffering from severe asthma.[5]

OBSERVATIONS OF ROBERT S. IVKER, D.O.

In his 2001 book, *Asthma Survival,* Dr. Ivker included more than 35 pages about the relationship of *Candida albicans* to asthma. In discussing his own observations in practice, Dr. Ivker said:

> . . . the overuse of antibiotics has contributed to a gradual weakening and dysfunction of the immune system that facilitates the survival

and growth of these supergerms, in addition to causing an overgrowth of yeast organisms, or candidiasis. In fact, a 1999 study performed in New Zealand at the Wellington School of Medicine, found that antibiotic use in infancy may be associated with a significantly increased risk of developing asthma.

In at least two-thirds of asthmatics, there is a history of recurrent sinusitis. Almost every one of these sinus infections is treated with broad-spectrum antibiotics. In the first Sinus Survival Study, completed in March 2000, the 10 participants each had moderate to severe chronic sinusitis that had not responded to conventional treatment—antibiotics and, in several instances, surgery. All of them scored above 180 on the Candida Questionnaire, four had asthma besides the sinusitis, and each was treated with the antifungal (candida is a type of fungus) drug Diflucan 200 mg./day for six weeks in addition to the entire Sinus Survival Program.

Following this course of treatment, nine of the 10 participants experienced a dramatic improvement in their sinus symptoms (many were feeling better than they had in years), while three of the four asthmatics reported similar results with their asthma—much less difficulty breathing, along with a significant reduction in their need for their inhalers . . . This study provides evidence supporting the findings of the 1999 Mayo Clinic study. They reported that *an immune system response to fungus rather than bacterial infection is the cause of most cases of chronic sinusitis.*

SUCCESS STORIES

Tabitha Vernier*

In February 2001 I received an e-mail from Kathy Vernier of Virginia. She told me about her daughter's long struggle with respiratory problems. Here are excerpts:

Tabitha has been troubled by sinus problems, bronchitis and recurrent bouts of asthma since early childhood . . . She's also been hospitalized several times. Then three years ago, we took her to Dr.

---

*You'll find Tabitha's story on pages 144–146 of *Yeast Connection Success Stories.*

George Ward, a physician in the allergy department at the University of Virginia . . . Although Tabitha gave no history of skin fungal infections, he prescribed Diflucan and it proved to be a wonder drug, enabling her to overcome her asthma and lead a normal life.

To get more information, I called Tabitha in May 2001 and with her permission, I recorded our conversation. Here are excerpts of the transcript:

My family doctor continues to give me Diflucan two or three times a week. It seems like I'd have bronchitis two or three times a year . . . It would be treated with some sort of antibiotic. That would take care of the infection, but the cough would stay. I'd end up with a cough that would keep me up all night . . . It just wouldn't go away . . . That's how I ended up at UVA for allergy tests. The only test I showed positive to was house dust mites. Then I was given Diflucan, and within a week, the cough was gone and I'd had it for about three months.

In a message I received from Tabitha's mother, Kathy, in the summer of 2002, she said, "Tabitha has moved to Oregon and is doing well and has had no recurrence of her asthma. No lapse of her asthma or bronchitis."

Bolin Stumb

In 1989, I saw Bolin because of repeated health problems that included fatigue, skin and respiratory problems. Here are excerpts from a letter she sent me in the summer of 2000:

When I was 13, I had my first asthma attack. Persistent respiratory problems, as well as recurrent episodes of bronchitis troubled me for many years. Trying to find help I saw several allergy doctors in Nashville and underwent allergy work-ups, including skin tests. I took allergy shots from these doctors for a total of about 12 years. My health did not improve. In fact, I continued to get worse.

During the mid-1980s I took many antibiotics and a lot of prednisone. I was hospitalized several times and given respiratory therapy. I suffered from severe coughing attacks for many years and, in the late 1980s, I broke my ribs during one of these attacks. As you know, I was pretty discouraged because of all of these problems, but after you put me on the treatment program which included dietary changes, antifungal medication and CoQ$_{10}$, I was healthy for the first time that I could remember.

To get an update I called Bolin in September 2002 and she said:

I'm continuing to do well! I take a maintenance dose of Diflucan twice a week. I tried cutting it down to once a week and I began to show mild symptoms, so I'm back on the twice a week dose.

MY COMMENTS

Asthma, like many other chronic disorders, develops from multiple causes. Yet, I feel that any person with chronic asthma who has received repeated antibiotic drugs and/or corticosteroids should be given a trial of oral antifungal medications, probiotics and a sugar-free diet.

I find that food and other sensitivities usually improve when the adrenal insufficiency and yeast or parasitic overgrowth are treated.

REFERENCES

1. van der Brempt, X., Mairesse, M., and Ledent, C., "Ketoconazole (K) in Asthma: A pilot study," *J. Allergy Clin. Immunol.*, January 1994 (abstract).

2. Ward, G. W. et al, "Treatment of late-onset asthma with fluconazole," *J. Allergy Clin. Immunol.*, 1999; 104:541–6.

3. Ward, G. W., Hayden, M. L., Rose, G., Call, R. C., Platts-Mills, T., "Trichophyton Asthma: Reduction of Specific Bronchial Hyperreactivity Following Long Term Antifungal Therapy," *J. Allergy Clin. Immunol.*, January 1994 (abstract).

4. Nsouli, T. L., Annual Meeting of the American Academy of Allergy, Asthma and Immunology, Washington, D.C., 1999.

5. Itkin, I. H. and Dennis, M., "Bronchial hypersensitivity to extract of Candida albicans," *Journal of Allergy*, 1966; 37:187–195.

CHAPTER 25

Sinusitis

Sinusitis follows on the heels of asthma and allergies. In fact, it's closely connected and, I suspect, yeast plays just as important a part in the aggravating symptoms of sinusitis.

Almost 32 million Americans suffer from chronic sinusitis. That's 16.3% of the population and I was surprised to learn the majority of the sufferers are women.

I've been interested in sinusitis since my medical school days. Although I enjoyed excellent health, I was bothered by a stuffy nose and mucus in the back of my throat. Although I tried various therapies, I was bothered by these problems until I visited internist/allergist Theron Randolph in Chicago in the mid-1950s.

During our visit, Dr. Randolph noted that I was rubbing my nose and clearing my throat. When he learned that cow's milk was my favorite beverage he said, "Dr. Crook, milk could be causing your nasal congestion and sinus problems." I stopped drinking milk, and my sinus symptoms went away. Since then, they've only returned when I consume too many dairy products.

Over the years, I've found that many of my patients were helped by removing milk and other dairy products from their diets.

In the 1980s, after learning about *Candida albicans* from Dr. C. Orian Truss, I prescribed oral antifungal agents for many of my patients with nasal congestion and sinusitis, and some of them were helped by sniffing powdered nystatin.

During a visit to Washington, D.C. in the early 1990s, I had lunch with Dr. Alexander Chester, a board-certified internist. I was delighted and excited to know of his interest in food allergies and sensitivities and chronic fatigue syndrome. Here are excerpts from his letter published in the spring 1995 issue of the *CFIDS Chronicle*.

> Most chronic fatigue is treated by internists, who . . . have received little information about sinusitis in their training. Large internal medicine and medical specialty techs often devote little more than a paragraph to chronic sinusitis.

I became more interested in the yeast connection to sinusitis after I read a 1999 Mayo Clinic report by otolaryngologist Jens U. Ponikau and colleagues.[1] In a research study of more than 200 consecutive patients with chronic rhinosinusitis, these investigators found fungi in the sinuses of 96% of the patients and in over 20%, the fungus was *Candida albicans*.

While Dr. Ponikau found fungi both in people with and those free of chronic sinusitis, it is only those with sinusitis who have an allergic reaction to the fungi. For this condition, Dr. Ponikau recommends using antifungal nasal sprays. Results have been so dramatic using this treatment that the Mayo Clinic has applied for a patent on the spray.

Then in early 2000, I read the fourth edition of Dr. Robert Ivker's book, *Sinus Survival*.[2] I was delighted to see that many pages of this book focused on *Candida albicans*. Here are excerpts:

> I'm thrilled that medical science has now found objective evidence supporting the treatment of most cases of Type I (most severe) chronic sinusitis with antifungal medication . . . In March 2000, in collaboration with William Silvers, M.D., a Denver allergist, the first Sinus Survival Study was completed. Each of the participants was a patient of Dr. Silvers with a long-term history of moderate to severe (Types I and II) chronic sinusitis. Every one of these patients scored above 180 on the Candida Questionnaire and Score Sheet, and each was treated with Diflucan, a powerful antifungal drug, in addition to the rest of the Sinus Survival Program.
>
> After four months on the program, including six weeks on Diflucan, all but one of the participants . . . experienced a very significant improvement in their condition. The majority reported feeling better than they had in years . . .
>
> Any medication that can potentially cause gastrointestinal ulcer-

ations or inflammation and weaken the lining of the gut can allow candida to gain a stronger and deeper foothold . . . While candida thrives on it, sugar weakens our immune system. It decreases the ability of white blood cells, phagocytes in particular, to engulf unwanted organisms . . .

Typically in patients with chronic sinusitis, the primary causes are: (1) repeated broad-spectrum antibiotics, along with (2) a sugar-filled diet and (3) significant emotional stress. As a general rule in medicine, as in life, there is rarely just one cause for anything. *However, in my experience, in almost every instance of a particularly resistant case of chronic sinusitis, candida is a primary cause.*

In a September 2002 letter to me, Dr. David Morris, M.D., a La-crosse, Wisconsin allergist, offered comprehensive diagnostic and treatment options for the allergic patient:

I'm very proud and pleased with the number of people I've been able to help with chronic sinus problems—probably several thousand patients over the last 10 years . . . These people are very grateful to get help . . .

We know from the research at the Mayo Clinic that over 90% of people with chronic sinus problems showed fungi in their nose and sinuses, and in more than 20%, it was candida.

Dr. Morris, a fellow of the American College of Allergy and Immunology and a Diplomate of the American Board of Allergy and Immunology, carried out pioneer work using sublingual therapy for people with food and chemical sensitivity.

He said that he prescribed Diflucan in helping many of these patients, and if they were more allergic to aspergillus than to candida, he used Sporanox, two or three capsules, twice weekly, with food.

In commenting on sinusitis, Charles W. Cox, M.D., a recently retired otolaryngologist in Jackson, Tennessee, said:

The gold standard for treatment of chronic sinusitis that has failed to clear with appropriate antibiotic therapy, allergy and environmental control measures remains surgical. By relieving the obstruction to sinus drainage pathways, the majority of cases will be cured.

In discussing allergic fungal sinusitis, Dr. Cox said that it had been poorly understood until the recently published studies by Mayo Clinic otolaryngologists that showed a high incidence of fungal organisms. Fungal sinusitis is primarily a defect in the immune system, and long-term broad-spectrum antibiotics predispose people to develop fungi in their sinuses.

In his continuing discussion he said that more and more otolaryngologists and other physicians are using topical antifungal medications and some are using oral antifungal medications. Controlling the environment by removing fungal contamination and using environmentally safe methods may be the key to the long-term control of fungal sinusitis.

REFERENCES

1. Ponikau, J. U., et al., "The Diagnosis and Incidence of Allergic Fungal Sinusitis," *Mayo Clinic Proc.*, 1999, 74:877–884.

2. Ivker, R. S., *Sinus Survival,* 4th Edition, Penguin Putnam, New York, 2000.

PART 5

Steps You'll Need to Take to Regain Your Health

CHAPTER 26

Dr. Crook's 10-Step Program to Regain Your Health

If this book can give you just one thing, I hope it is the realization that:

You can overcome your health problems and get your life back on track!

Believing this simple statement is your crucial first step. Of course, you'll need help, including professionals and non-professionals. You'll need love and support from family members, friends, and support groups.

> ### Make this your motto:
>
> "If it's going to be, it's up to me."

No one says this will be easy, but your life hasn't been easy up to this point. If your health problems are yeast-connected, a sugar-free diet (and possibly a yeast-free diet) plus antifungal medication will start you on the road to recovery. You'll have to add in a large dose of patience. It may take months or even a year or more.

We'll go into this program in much greater detail later in the book, but here are the bare bones of what you'll need to do to get well:

1. Clean Up Your Diet

Go to your kitchen and get rid of the sugar, corn syrup, white bread and other white-flour products, soft drinks and most ready-to-eat cereals. Foods and beverages containing these nutritionally deficient simple carbohydrates promote poor health and feed the candida yeast.

You'll also need to get rid of hydrogenated and partially hydrogenated oils and replace them with modest amounts of unrefined oils, including flaxseed, olive and coconut.

Avoid yeasty foods and beverages, especially dried fruits, mushrooms, alcoholic beverages, all leavened breads, bagels, pastries, pretzels, pizza and rolls. In two or three weeks, after you've gotten a bit better, you can gradually try these foods again and see if they bother you.

Replace these empty foods with more vegetables, healthy oils and some of the "grain alternatives," quinoa, amaranth and buckwheat. I'll tell you more about those later.

See www.yeastconnection.com for a printable grocery list of recommended foods to help you make changing your diet easier.

I know. This isn't easy, but it's not forever. After a few weeks or months, you may be able to relax a bit. Yet, you must stick to your diet until you show significant improvement. This is the essential component of your health plan. Without it, you won't get well.

2. Control Chemical Exposures

Almost without exception, people with yeast-related health problems are sensitive to the chemicals they encounter in everyday life. These include tobacco smoke, perfumes, colognes, glues, carpet and fabric odors, paints, formaldehyde, insecticides, diesel fumes and other traffic odors.

Although you can't avoid all of these, you can clean up your home. By getting rid of odorous bathroom and kitchen chemicals and insecticides, you can immediately lighten your load and your symptoms will start improving.

It's also a good idea to do what you can to clean up your workplace.

You may also react to chemicals in and on your food—even fresh foods, as well as those in cans, plastic wrap and other food containers. Where possible, buy and eat organic foods, and choose prepared foods in glass containers.

3. Lifestyle Changes

You'll need fresh air, sunlight, exercise and a proper amount of sleep. Don't be a couch potato spending endless hours in front of the TV or your computer screen. Take a walk. Do calisthenics. Play outdoor games. Move!

 See **www.yeastconnection.com** for a printable daily time diary to help you recognize where you might need to make some lifestyle changes.

4. Support

You'll need emotional and psychological nutrients, whether you are sick or well. Here are a few of them:

- Love
- Encouragement
- Praise
- Touch
- Hugs
- Laughter

They'll strengthen your immune system and play an important role in your renewed health.

 See **www.yeastconnection.com** to sign up through our online weekly e-mail service offering encouraging words you can share with others.

5. Nutritional Supplements

Everyone needs a good regimen of vitamins, minerals and other supplements.

Be sure you get supplements that are yeast-free, sugar-free and color-free.

Choose:

- A good multivitamin with minerals like calcium, magnesium and zinc, plus antioxidants like vitamins C and E, along with digestive enzymes.
- Essential fatty acids (EFAs), including flaxseed oil or ground flaxseeds, an excellent source of Omega-3 fatty acids and Omegfa-6 fatty acids such as evening primrose oil, borage and black currant seed oils.

6. Non-prescription Anticandida Supplements*

After you've completed the first four steps, you're ready to add supplements that help control the candida overgrowth:

- Probiotics: These preparations of lactobacillus acidophilus and other friendly probiotic bacteria (literally meaning pro-life, rather than antibiotic or anti-life) help crowd out candida in your digestive tract. Usual dose: ¼ to ½ tsp. of powder or 1–2 capsules, one-to-four times daily.
- Citrus seed extracts: Made from grapefruit seed. These supplements are designed to help take out "unfriendlies" in your intestines. Usual dose: 1–2 capsules or 2–6 drops of liquid in water, one-to-three times daily. There is some evidence these may be as effective as some prescription antifungals. Caution: You should not use citrus seed extracts if you are taking medication to lower cholesterol or calcium channel blockers for hypertension, congestive heart failure or migraines. If you're in doubt, check with your doctor.
- Garlic: Eat lots of garlic to keep those intestinal flora under control. If you can't stomach fresh garlic or you're losing friends because of garlic breath, try Kyolic deodorized capsules. Usual dose is two 300-mg.capsules twice a day, with meals.

---

*More about these in Chapter 29.

- Caprylic acid: A fatty acid derived from coconuts, it's been shown to kill fungus on contact. Usual dose: 1–2 capsules with each meal.
- Mathake: This tropical herb's name means "furry white tongue" in Fijian—an apt description of a major symptom of candidiasis. It's available as a tea or extract.
- Tanalbit: This product is effective against harmful bacteria and fungi in the digestive system without attacking friendly organisms and is able to address problems in the colon.
- Olive leaf extract: Lab studies show that this powerful antimicrobial can eliminate everything from bacteria to yeasts, fungi and molds. Usual dose: 1–2 capsules three times daily between meals.

7. Prescription Medications*

Oral prescription antifungals are available in the United States and are safe and effective, although they are expensive, ranging from around $7 a pill for Diflucan and Sporanox and about $3.75 per pill for Nizoral to about 56 cents a pill for nystatin. Fortunately, many medical insurance plans will pay the cost if the diagnosis is vaginal yeast infection or oral thrush.

These safe and effective prescription drugs are at the hub of the pharmacological treatment for systemic yeast overgrowth:

- Diflucan (fluconazole): This systemic medication has been in use for nearly 30 years, and in the past 10 or 12 years, licensed for use by patients with severe immunosuppression, including cancer and AIDS. Since the early 1990s, physicians have also found it is useful for the treatment of recurrent vaginitis, fatigue and depression.
- Sporanox (itraconazole): This antifungal drug has been used for a decade and is an effective medication for people with a variety of fungal infections, although it can have serious side effects.
- Nizoral (ketoconazole): Another antifungal available since 1981 is one of the earlier forms of the azole family of antifungals. There are some rare side effects related to Nizoral, but it is less expensive than later generations of the drug.

---

*More about these in Chapter 28.

- Nystatin: This medication has been used for more than 40 years, and it's the safest medication listed in the Physician's Desk Reference. It's not water-soluble, and not absorbed into the bloodstream, so that means it doesn't reach candida in the respiratory tract or the deeper layers of the mucous membranes of the vagina or intestinal tract.
- Miconazole: This generic antifungal is the main ingredient in Monistat, the anti-yeast cream sold over the counter in the United States. While the cream would not be effective against systemic candida overgrowth, miconazole gel is available under the brand name Daktarin in Europe, Australia and Canada. Some of my colleagues say it is an excellent affordable alternative to the expensive azole drugs available in the United States.

8. Work with a Kind and Caring Health Care Professional

Although the first five steps may help you improve, ideally your treatment program should be directed by a knowledgeable physician. Moreover, you can only get the anti-yeast medications Diflucan, Sporanox, Nizoral or nystatin from an M.D., and you'll need to be monitored as you're taking it.

Take the steps outlined in Chapter 9 to find a professional willing to work with you who really cares about your improvement.

 See **www.yeastconnection.com** for a printable packet of guidelines for talking with your physician.

9. Track Down Hidden Food Sensitivities

If you've faithfully executed steps 1–7 and you continue to experience significant problems, you may be sensitive to foods you're eating every day. Common troublemakers include yeast, milk, wheat, corn, eggs and citrus.

10. Faith, Courage, Persistence and Determination

These are key ingredients in any treatment program.

CHAPTER 27

Diet

Here it comes: the special sugar-free, anti-yeast diet I've mentioned so many times. It is the centerpiece of the entire anti-yeast plan.

I don't mean to sound harsh here, but it's important that you understand this: Without the diet, you won't get the results you are hoping for. You can take the medications and the supplements, and they may or may not help. You may feel better for a while, and then find your symptoms have returned.

> If your health problems are yeast-connected, you must eliminate sugar from your diet for at least three weeks and probably indefinitely.

It's not easy, but once you start feeling so much better, you'll be willing, even eager, to make these sacrifices.

The plan hinges on complete elimination of sugar and yeast from your diet for at least three weeks, and then you'll become your own detective by gradually reintroducing specific foods and recording your results.

 See www.yeastconnection.com for options designed to support your successful change to a sugar-free/yeast-free diet.

You'll want to keep a diary to jot down your progress and the results you get from your detective work.

I like to think of this as a four-step process:

First Step: Elimination—This is unquestionably the hardest part. You will eliminate all sugar and foods containing sugar, processed foods that often contain sugar and chemical additives, fruit, and all yeasts and fungi for the first three weeks.

Second Step: Challenge—Now that you're feeling much better and you've probably eliminated a large amount of yeast overgrowth from your digestive tract, it's time to challenge yourself, by experimenting with complex sugars, fruits and yeast to see what affects you. It'll feel a little like a backward step, but it's important to understand what foods trigger your symptoms so you can avoid them.

Third Step: Reassessment—If after a month or two on the anti-yeast program, you're still experiencing problems, you'll need to track down your hidden food allergies to further refine what you found out about your food sensitivities. (See Chapter 23 of this book and, for more details, read my book, *Tracking Down Hidden Food Allergies,* 1978, now in its 10th printing.)

Fourth Step: Maintenance—Now that you know your triggers, you can lighten up a little on the program. You can fine-tune your personalized program because you know what works. If you've passed the fruit challenge, you can add moderate amounts of fruit to your diet, probably not more than three or four pieces a week. You might be able to tolerate bread a couple of times a week without triggering symptoms. A piece of birthday cake may not send you into an immediate tailspin, if you've been very careful.

Now for the details:

STEP 1: ELIMINATION

Prepare your supplies and reorganize your kitchen and your way of thinking about food.

Think about what you'd need to do if you were taking a three-week ocean voyage on a sailing ship. Before starting your trip, you'd have to get enough food. I'm not asking you to buy all of it and store it in your pantry and refrigerator, yet here are my suggestions for getting started.

Clean out your cupboards:

- Go through your kitchen, pantry and refrigerator and get rid of the sugar, corn syrup, white bread and other white-flour products, soft drinks, most ready-to-eat cereals and all the sweet, fat-laden snack foods. Foods and beverages containing these nutritionally deficient simple carbohydrates encourage yeast overgrowth and promote poor health. To overcome your candida-related health problems, you'll need to avoid them.
- Replace them with more vegetables of all kinds, including those you may not usually eat. Also, go to your health food store and buy grain alternatives, including amaranth, buckwheat and quinoa. (You'll find instructions and recipes for preparing and serving them in *The Yeast Connection Cookbook.*)
- Get rid of processed and prepared junk foods, which have hydrogenated or partially hydrogenated fats, as well as those containing food coloring and additives. Add modest amounts of olive, walnut, flaxseed, sesame and other unprocessed, unrefined oils.
- Shop mainly around the outer edges of your supermarket. Look for fresh and frozen vegetables, fresh meat, poultry, seafood, eggs, tofu, olive oil, pure butter and sardines packed in sardine oil. I especially recommend organically grown foods which haven't been chemically contaminated. You'll find these foods in many health food stores and in some supermarkets.
- You'll also need to eliminate fruit for these first three weeks. In general, fruit is good for you, but in your condition, it's simply too high in sugar when you're fighting yeast overgrowth.

What You Can Eat During the First Three Weeks

Foods You Can Eat Freely—Low-carbohydrate vegetables. These vegetables contain lots of fiber and wonderful essential nutrients. They are relatively low in carbohydrates and calories. You can eat them fresh or frozen, cooked or raw.

- Asparagus
- Beet greens
- Bell peppers
- Broccoli
- Brussels sprouts

- Collard greens
- Cucumbers
- Daikon
- Dandelion
- Eggplant

- Cabbage
- Carrots
- Cauliflower
- Kale
- Kohlrabi
- Leeks
- Lettuce (all varieties)
- Mustard greens
- Okra
- Onions
- Parsley
- Parsnips
- Celery
- Endive
- Garlic
- Radishes
- Rutabaga
- Shallots
- Snow peas
- Soybeans
- Spinach
- String beans
- Swiss chard
- Tomatoes, fresh
- Turnips

Meat, Seafood, Eggs and Other Food

- Beef, lean cuts
- Chicken
- Cod
- Lamb
- Mackerel
- Other fresh or frozen fish
- Pork, lean cuts
- Salmon
- Sardines
- Shellfish: shrimp, lobster, crab
- Tofu
- Tuna
- Turkey
- Veal
- Wild game

Nuts, Seeds and Oils (unprocessed)

- Almonds
- Brazil nuts
- Cashews
- Filberts
- Flaxseeds
- Pecans
- Pumpkin seeds
- Butter (in moderation)
- Oils, cold-pressed and unrefined: Corn, Olive, Safflower, Soy, Sunflower, Walnut

Foods You Can Eat Cautiously

High-Carbohydrate Vegetables

- Artichoke
- Avocado
- Beans, peas and other legumes
- Celery root (celeriac)
- Fennel
- Potatoes, sweet
- Potatoes, white

- Beets
- Boniata (white sweet potato)
- Breadfruit

- Winter, acorn or butternut squash

Dairy Products*

- Cream cheese
- Hard cheeses
- Yogurt

Whole Grains

- Barley
- Corn
- Kamut
- Millet
- Oats

- Rice
- Spelt
- Teff
- Wheat

Grain Alternatives

- Amaranth
- Buckwheat
- Quinoa

Breads, Biscuits and Muffins—All breads, biscuits and muffins should be made with baking powder or baking soda as a leavening agent. You'll find recipes and more information in *The Yeast Connection Cookbook*. Do not use yeast unless you pass the yeast challenge as described on page 201.

Foods You Must Avoid Completely

Sugar and foods containing sugar. Avoid sugar and other quick-acting carbohydrates, including sucrose, fructose, maltose, lactose, glycogen, glucose, mannitol, sorbitol, galactose, monosaccharides and polysaccharides. Also avoid honey, molasses, maple sugar, date sugar and turbinado sugar.

Packaged and processed foods. Canned, bottled, boxed and other

---

*Eat these sparingly since so many people have food sensitivities to dairy products. If your symptoms persist, eliminate them completely.

packaged and processed foods usually contain refined sugar products and other hidden ingredients.

You'll not only need to avoid these sugar-containing foods in the early weeks of your diet, but you'll probably need to avoid them indefinitely.

Avoid yeast-containing foods the first 10 days of your diet. Here's a list of foods that contain yeasts or molds:

- Breads, pastries and other raised-bakery goods.
- Cheeses: All cheeses. Moldy cheeses, such as Roquefort, are the worst.
- Condiments, sauces and vinegar-containing foods: Mustard, ketchup, Worcestershire, Accent (monosodium glutamate); steak, barbecue, chili, shrimp and soy sauces; pickles, pickled vegetables, relishes, green olives, sauerkraut, horseradish, mincemeat and tamari. Vinegar and all kinds of vinegar-containing foods, such as mayonnaise and salad dressing. (Freshly squeezed lemon juice may be used as a substitute for vinegar in salad dressings prepared with unprocessed vegetable oil.)
- Malt products: Malted milk drinks, cereals and candy. (Malt is a sprouted grain that is kiln-dried and used in the preparation of many processed foods and beverages.)
- Processed and smoked meats: Pickled and smoked meats and fish, including bacon, ham, sausages, hot dogs, corned beef, pastrami and pickled tongue.
- Edible fungi: All types of mushrooms, morels and truffles.
- Melons: Watermelon, honeydew and, especially, cantaloupe.
- Dried and candied fruits: Raisins, apricots, dates, prunes, figs and pineapple.
- Leftovers: Molds grow in leftover food unless it's properly refrigerated. Freezing is better.

What You Should and Should Not Drink

Water: You should drink eight glasses of water a day. Yet, ordinary tap water may be contaminated with lead, bacteria or parasites.

Fruit Juices: These popular beverages are a big "no-no," even more so than eating fresh fruit. Most fruit juices, including frozen, bottled or canned, are prepared from fruits that have been allowed to stand in bins, barrels and other containers for periods ranging from an hour to several days or weeks. Although juice processors discard fruits that are obviously spoiled by mold, most fruits used for juice contain some level of mold.

Coffee and Tea: These popular beverages, including the health food teas, are prepared from plant products. Although these products are subject to mold contamination, most people seem to tolerate them. To decide, you can experiment. Teas of various kinds, including taheebo (Pau d'Arco), have been reported to have therapeutic value. If you can't get along without your coffee, limit your intake to one or two cups a day. Drink it plain or sweetened with stevia.

Alcoholic Beverages: Wines, beers and other alcoholic beverages contain high levels of yeast contamination, so if you're allergic to yeast, you'll need to avoid them. You should stay away from alcoholic beverages for another reason: They contain large amounts of quick-acting carbohydrate. If you drink these beverages, you'll be feeding your yeast.

Diet Drinks: These beverages possess no nutritional value. Moreover, they're usually sweetened with aspartame (Nutrasweet), which causes adverse reactions in many people. They also may contain caffeine, food coloring, phosphates and other ingredients that many individuals can't tolerate. However, since diet drinks do not contain mold, some people with candida-related problems can tolerate them. If you drink them, use them sparingly.

Meals

The menus listed on the next few pages are all sugar-free and yeast-free and designed to help you answer that always troublesome question, "What can my family and I eat?"

The menus for the early weeks are also fruit-free and contain relatively few grains and high-carbohydrate vegetables (such as potatoes, yams and lima beans). Depending on your likes and dislikes, using these general guidelines, you can change these menus to suit your tastes and those of other members of your family.

Breakfasts
- Oatmeal with butter or flaxseed oil and pecans.
- Brown rice with filberts, sardines packed in sardine oil and rice cakes.
- Well-cooked eggs, two strips of crisp bacon and grits with butter.
- Oatmeal, pork chops and cashew nuts.
- Cooked amaranth, tuna, tomatoes and walnuts.
- Cooked quinoa, baked sweet potato and pecans.

Lunches
- Wraps made with romaine leaves: roll-ups with turkey or tuna, grated carrots.
- Caesar salad with grilled chicken breast. (Use a small amount of lemon juice instead of vinegar in the dressing.)
- Kitchen sink salad: every veggie you can find on a bed of romaine lettuce, perhaps with a few julienne strips of turkey.
- Unsweetened yogurt and a handful of nuts.
- Hard-boiled eggs and carrot and celery sticks.
- Scoop of egg, tuna or chicken salad (made with soy mayonnaise) and a tomato.
- Hamburger patty, lettuce, tomato and onions.
- Easy veggie soup (Progresso makes a decent canned soup) with rice cakes.

Suppers
- Baked Cornish hen, steamed cabbage, asparagus, salad with lettuce and pecans. Use walnut oil and lime juice dressing.
- Steak (or hamburger patty), eggplant, mixed green salad with cucumbers and green peppers.
- Pork chops or lamb chops, turnip greens, okra, carrot and celery sticks.
- Roast turkey, baked acorn squash, steamed spinach, grated cabbage, almonds with lemon juice and flaxseed oil dressing.
- Mixed vegetables cooked in microwave, pecans.
- Tuna fish, broccoli, black-eyed peas.

Warning: You May Experience Die-Off

If you have massive candida overgrowth, you may experience a condition called "die-off." This means millions of candida yeast cells are dying off in your digestive tract and perhaps in other parts of your body, especially if you have begun taking antifungal medications. When large numbers of yeasts are killed, metabolic products are released. Until your body gets rid of them, symptoms may continue or even increase.

> Symptoms include fatigue, depression, aches, irritability and abdominal pain.
>
> For a few days, you may feel much worse, and this is the stage at which many people decide this plan is not for them.
>
> Please don't give up! Remember, you are doing the best possible thing for your body. If you are in a position to take it easy for the few days of the die-off, so much the better. Drink lots of water and know that you are making progress!

Die-off symptoms can be lessened if you start the diet a week or two before you start taking antifungal medications. You can also take 3–5 grams of vitamin C daily plus 400 IU of vitamin E three times a day and special enzymes taken between meals to help eliminate the toxins.

STEP 2: CHALLENGE

Now is the time to figure out what's playing havoc with your body. If you've been faithful in Step 1 (the Elimination phase), you've made a great start, and your yeast overgrowth is beginning to get under control.

Now it's time to challenge your system and see what you can and cannot tolerate.

It's really up to you how long this step takes, but remember: You must remain on the Step 1 diet throughout this stage with the exception of the foods you are reintroducing as part of your research.

First, take the yeast challenge:

After you've avoided yeast-containing foods for three weeks, you can find out if you're sensitive to yeast by eating a tablet of brewer's yeast, which you can find at a health food store.

If it doesn't bother you, eat a couple tiny bites of moldy cheese, such as blue cheese or brie the following day.

Try a couple of tablespoons of vinegar, perhaps mixed into your salad dressing.

Record your responses to these foods over several days. It's important not to add more than one thing at a time so you can be sure you've really tracked down the culprits.

If consuming these yeasty foods triggers symptoms, stay away from them for several weeks. Then you can experiment further. Your sensitivity may diminish as you eliminate the yeast overgrowth.

Truly yeast-free diets are impossible to come by because you'll find yeast and molds on the surfaces of all fruits, vegetables and grains. Once you've discovered that you're sensitive to yeast, you'll need to be your own judge as to how well you tolerate food that may contain some yeasts or molds.

Now take the fruit challenge:

You have avoided all fruits during the first three weeks of your special diet. Now, to see if they bother you, you can do the fruit challenge. Here's how:

Take a small bite of a banana. Ten minutes later, eat a second bite. If you have no reaction in the next hour, eat the whole banana.

If you tolerate the banana without developing symptoms, try strawberries, pineapple, apple or citrus fruits the next day. If you show no symptoms following these fruit challenges, chances are you can eat fruit in moderation. But feel your way along and don't overdo it.

 See www.yeastconnection.com for a printable chart of the most nutritious fruits and vegetables.

Chances are you will find you can eat freely . . .

- all fresh vegetables
- all fresh fruits (in moderation)
- whole grains (in moderation)

You can continue to consume fish, lean meat, eggs, nuts, seeds and oils. And if you pass the yeast-challenge test, you also can include some of the yeast-containing foods.

You must continue to avoid . . .

Sugar, maple syrup, honey, corn sugar, date sugar and sugar-containing foods, packaged and processed foods of low nutritional quality that contain sugar, and hydrogenated or partially hydrogenated fats and oils.

You may find you feel "mad at the world" because you aren't getting the foods you crave. You may act like a two-pack-a-day smoker who quit smoking "cold turkey." Here's why: People who suffer from hidden food allergies are often addicted to the foods causing their problems.

STEP 3: REASSESSMENT

You should also rotate other foods, especially during the early weeks and months of your treatment program. Here's why:

Many, and perhaps most, individuals with yeast-connected health problems are allergic to several (and sometimes many) different foods. The more frequently you eat a particular food, the greater your chances of developing a "hidden" allergy to that food. These allergies may contribute to your fatigue, headaches, muscle aches, depression or other symptoms. And strange as it may seem, you may become addicted to the foods that are causing your symptoms, so you crave them.

 Remember to see www.yeastconnection.com for printable "Food Record and Symptom" forms to help you explore your sensitivities.

You may find you can add in a few foods without dire consequences. Now is the time to determine those food allergies and sensitivities, especially if you're not yet feeling as good as you would like.

You can eat and drink the following:

- any vegetable but corn
- any fruit but citrus

- any meat but bacon, sausage, hot dogs or luncheon meats
- rice, oats and the grain alternatives amaranth, quinoa and buckwheat
- unprocessed nuts that are fresh, refrigerated or frozen
- water, preferably bottled or filtered

Avoid these:

- chocolate
- citrus
- corn
- food coloring and additives
- fruit punches
- milk
- processed and packaged foods
- soft drinks
- sugar
- wheat
- yeast

When you add a food to your diet, be sure to add it in pure form. For example, when you add wheat, use pure whole wheat rather than bread, since bread contains milk and other ingredients that may confuse your results. If you're testing milk, use whole milk rather than ice cream, since ice cream contains sugar, corn syrup and other ingredients.

If you identify one or more diet troublemakers and are still having problems, try the Cave Man Diet—which consists solely of protein and vegetables.

Eliminate all the common allergy-causing foods in the "avoid" list above. In addition, you must avoid:

- beef
- chicken
- coffee
- eggs
- oats
- pork
- rice
- tea
- tomatoes
- wheat
- white potatoes
- any food or beverage you eat more than once a week

Now begin returning these foods to your diet one at a time, giving at least a day or two in between each to experience your results and identify your hidden food sensitivities.

Any food can be a troublemaker, so if you're still not feeling top notch after Step 3, search out some other possible allergens such as soy or shellfish.

Here's the good news: If the foods you've avoided are causing your symptoms, you'll usually feel better within four to six days of eliminating this food. Almost always, you'll improve by the 10th day. Occasionally, though, it'll take two or three weeks before your symptoms go away completely.

Rotating your foods will help you identify those that may be disagreeing with you and causing your symptoms.

In rotating your diet, you eat a food only once every four to seven days. For example, in rotating fruits, you'd eat oranges on Monday, bananas on Tuesday, apples on Wednesday and pineapple on Thursday. Then on Friday you could start over again with oranges (or a related food such as grapefruit). Do the same thing with other food groups, including meats, vegetables and grains.

This will take some time, but it's worth it!

STEP 4: MAINTENANCE

Congratulations! Now you've really done your homework and you know what you can and cannot eat.

No doubt, you are feeling vastly better than you felt just a few short weeks ago. Although this wasn't the primary purpose, you've likely dropped a few pounds, too. That's a welcome bonus for many of us!

You've also made the anti-yeast diet a part of your life.

So you can loosen your grip from time to time. Eat a slice of birthday cake on your birthday. Sip a glass of red wine once a week or so. You may be able to get away with it, or you may not. Continue to monitor your progress and if you have a setback, look at what you ate and take corrective measures.

Your doctor may keep you on antifungal medications indefinitely or you may discover that you need them from time to time if you slip off the anti-yeast diet.

Sadly, for most of my patients, simply eliminating the yeast overgrowth does not seem to "cure" the problem. It may be that the immune system is deeply affected, or it may be that you are particularly prone to imbalances in the digestive tract that promote yeast overgrowth.

Whatever the reason, it is usually necessary to remain on the maintenance diet indefinitely to maintain your newfound sense of well-being.

Every person is unique. In following the anti-candida diet, use a trial-and-error approach. Most of my patients with candida-related illnesses can follow a less-rigid diet as they improve, especially if they're following other measures to regain their health. You need to include nutritional supplements, exercise, stress reduction and avoid exposure to environmental chemicals and mold spores.

Meal Suggestions After Step 1

Breakfasts

- Ground beef patty, scrambled eggs, grits with butter, applesauce muffin.
- Pork chop, steamed Brussels sprouts, whole wheat biscuit, grapefruit.
- Toasted rice cakes with peanut butter, sliced banana, turkey burger.
- Brown rice with butter and chopped almonds, tuna (water-packed), fresh pineapple.
- Well-cooked eggs, any style; pancakes made with teff, spelt, kamut; freshly squeezed orange juice.
- Barley cereal with banana and pecans, milk, fish (baked or broiled).
- Hot oatmeal with cashews, milk and fresh or frozen peaches.

Lunches

- Chili, corn bread, coleslaw, orange.
- Tuna salad on lettuce, whole-grain flat bread, fresh pineapple.
- Fruit salad and yogurt.
- Chicken salad, rice cakes, apple.
- Chef salad, whole wheat crackers, apple.
- Hamburger patty, raw celery and carrots, whole wheat biscuits, pear.

Suppers

- Sauteed liver, lima beans, baked acorn squash, sliced tomato, banana oat cake.
- Broiled fish, cabbage and carrot slaw, wax beans, whole wheat popovers, baked banana.

- Broiled lamb chops, steamed cauliflower, steamed broccoli, boiled potatoes, baked apple.
- Rock Cornish hen, steamed carrots and peas, wild rice, rice crackers.
- Roast duck, kale, barley soup, sweet potato, steamed green beans, corn bread.
- Broiled steak, baked potato, lettuce, tomato, cucumber salad with freshly squeezed lemon juice and safflower or linseed oil dressing, mixed greens, fresh strawberries.
- Chicken and rice, steamed artichoke, turnip greens, corn bread, pear.

SHOPPING TIPS

- Use whole foods.
- Use fresh fruits and vegetables. Commercially canned products often contain yeasts and added sugar. Buy fresh organic vegetables when possible.
- Avoid foods labeled "enriched" if you're allergic to yeast.
- Since many, and perhaps most, canned, packaged and processed foods contain hidden ingredients, including sugar, dextrose and other carbohydrate products, avoid them.
- If you must use canned or packaged foods, read labels carefully.
- If buying frozen vegetables, select those without added sauces or ingredients.
- Avoid processed, smoked or cured meats, such as salami, wieners, bacon, sausage and hotdogs, since they often contain sugar, spices, yeast and other additives. These foods also are loaded with the wrong kind of fat.
- Avoid bottled, frozen and canned juices. If you want juice, buy fresh fruit and prepare your own.
- Buy nuts from a natural food store, and make sure they are fresh and not rancid or contaminated with molds. Store them in your refrigerator or freezer. Avoid peanuts if you're allergic to yeasts or molds.

- All commercial breads, cakes and crackers contain yeast. If you want yeast-free breads, you'll have to obtain them from a special bakery or bake your own. Word of caution: Many people with yeast-related problems react adversely to wheat. So, if you continue to experience symptoms, you may need to avoid breads and similar products. Hain and Chico San or Golden Harvest Rice Cakes contain no sugar or yeast. Most rice cakes contain no sugar.
- Use expeller-pressed vegetable oils, such as sunflower, safflower, flaxseed and corn. Flaxseed oil is a superb source of the important Omega 3 essential fatty acids.
To make salad dressing, combine the oil with fresh lemon juice to taste.
- Buy whole grains (barley, corn, kamut, millet, oats, rice, spelt, teff and wheat) from a natural food store. Grains can be an important ingredient of a nutritious breakfast. Barley, rice and other grains can also be used in various ways at other meals. Barley or rice casseroles are especially tasty.
- For convenience, you can get excellent bagged organic salads.

Eating Out—If you're like most people, you "live on the run" and eat foods away from home. What's the answer? Do the best you can. And during the early weeks and months of your candida-control program, you may need to do a lot of brown-bagging. And when you eat out, you'll need to make your selections carefully to avoid foods that trigger your symptoms.

When eating on the run, plan ahead. Don't wait until you're rushing off to work. Make sure that you can fill your "brown bag" with nutritious foods, including raw vegetables, nuts and rice cakes. You also may get some of the nutritious vegetable-based burgers from your health food store.

Vegetarian—Vegetarian or modified vegetarian diets may lessen your chances of developing osteoporosis, heart disease, diabetes and other degenerative diseases, which affect tens of millions of Americans who consume high-protein, high-fat diets.

Although I included meats in the menus for the early weeks, I urge you to improve the quality of your diet in the months ahead by eating less meat. Here's one reason: Animal foods are loaded with pesticide residues. So eat more plant foods, including vegetables, fruits and whole grains. If you eat meat, make it organic.

Tempeh is an excellent source of vegetable protein that can be prepared in a wide variety of appetizing ways ranging from stir-fries to barbecued tempeh steaks. It's a fermented soy product and the safest way to consume soy.

Dry Cereals—These cereals, which you'll find in your supermarket . . . even the best of them . . . have been processed and subjected to high heat. Accordingly, they're much less desirable than hot cereals you prepare at home made from whole grain. Moreover, most of these cereals are loaded with sugar and contain malt and added yeast-derived B vitamins. So if you're allergic to yeast, you'll need to avoid them.

If you like dry cereals, you can find some nutritious ones at a health food store. Many of them are organically produced, some contain a mixture of grains, and most are fruit-sweetened. (I like Health Valley cereals.)

If you purchase a dry cereal at your usual grocery store, I suggest sugar-free, yeast-free Shredded Wheat. Other cereals that may be suitable include Cheerios, Puffed Rice, Wheat Chex, Puffed Wheat, Post Toasties, Product 19, Kashi and Special K. All have less than 6 percent added sugar. I don't recommend them for the early weeks, but as you improve, you may be able to tolerate these cereals in limited amounts.

You can find menus and suggestions that will make carrying out your diet detective work much easier in *The Yeast Connection Cookbook,* which I co-authored with Majorie Hurt Jones, R.N. This book contains more than 225 recipes that will help you with your meal planning, and it has excellent recipes using grain alternatives.

CHAPTER 28

Prescription Antifungal Medications

Throughout this book, you've read over and over about the prescription antifungal medications and how essential they are to your program to overcome yeast-related problems.

The systemic azole drugs are the gold standard of treatment for fungal problems, and especially for the treatment of *Candida albicans* overgrowth.

They are:

- Diflucan (fluconazole)
- Nizoral (ketoconazole)
- Sporanox (itraconazole)

In addition, nystatin has proven its worth as an antifungal for 50 years and so has another lesser-known azole, miconazole. Miconazole is the active ingredient in Monistat, the cream used to treat vaginal yeast infections. Miconazole is also helpful when taken internally in capsule form, but it's not available in the United States in the form needed for systemic treatment.

In this chapter, we'll look at the prescription antifungals, their benefits and their side effects.

Before I go too far, I want to caution you: These antifungals are prescription drugs for a reason. On rare occasions, there are very serious side effects associated with some of them. You absolutely should be under a doctor's supervision if you are taking any of them. While you may be able to find them on the Internet from sources outside the United States that do not require a prescription, this could be very detrimental to your health.

The amount of time you'll need to take these drugs may vary. Many of my colleagues like to use them for four to six months, sometimes as long as a year. The minimum amount of time they're used is two months. Maintenance doses may be necessary for an indefinite period if symptoms keep reappearing.

THE AZOLES—NIZORAL, DIFLUCAN, SPORANOX

In an article published in the *New England Journal of Medicine* almost a decade ago, the authors said that all three of the azole drugs are active against *Candida albicans*. Here are excerpts:

The oral azole drugs represent a major advance in systemic antifungal therapies. Among the three, fluconazole has the most attractive pharmacologic profile . . . However, ketoconazole is less expensive than fluconazole and itraconazole, an especially important consideration for patients receiving long-term therapy.[1]

Diflucan (Fluconazole)

Diflucan is the most commonly used medication for short-term treatment of vaginitis and is the drug of choice for many physicians treating yeast overgrowth on a long-term basis.

Taken in pill form, Diflucan works systemically to penetrate tissues infested with Candida organisms.

It's been widely used in Europe for about 20 years, and available in the United States since the early 1990s. It has been used in 57 countries, and 9 million women have used it for vaginal yeast infections. Initially, Diflucan was licensed for use by AIDS and cancer patients who were severely immunocompromised, but in 1993 it was approved for use in treating vaginitis.

During the past decade, many of my colleagues have told me they consider Diflucan an effective and safe medication that can be given for many weeks, months or even years.

Pfizer, the company that manufactures Diflucan, has noted a variety of side effects with the drug: The most common side effects are headaches (in 13% who took it in a clinical trial), nausea (7%), vomiting, abdominal pain and diarrhea. There have also been reports of reversible hair loss.

Pfizer warns: The most serious, but rare, potential side effect of Diflucan is liver toxicity. Symptoms of liver toxicity are unusual fatigue, loss of appetite, nausea, vomiting, jaundice, dark urine or pale stools.

The down side of Diflucan, like many antifungals, is its cost, somewhere around $7 a pill. It's available in 100-, 150- and 200-mg. doses, and while most doctors prefer 600 mg. a day or more at the outset of treating yeast overgrowth, once the overgrowth is under control, dosages can usually taper off to a maintenance dose as low as one pill a week. Most insurance companies cover Diflucan.

Free supplies of Diflucan may be available from Pfizer for those who have extremely limited financial resources and no insurance. To obtain further information, have your physician call Roerig Division, Pfizer, Inc., (800) 869–9979.

Sporanox (Itraconazole)

Sporanox is a "cousin" of Diflucan, approved by the FDA for use against thrush (candida infections in the mouth and throat).

Available in the United States since 1993, Sporanox is highly successful in eliminating the symptoms of thrush within seven to 14 days and has been clinically proven effective in some patients who are resistant to Diflucan, usually at a level of about 100 mg. once daily. Higher dosages (about 200 mg. a day) are usually prescribed for patients with persistent vaginal yeast infections.

Charles Resseger, D.O., of Norwalk, Ohio, told me he finds that

Diflucan and Sporanox work synergistically and are often very effective in patients who don't respond to Diflucan alone.

Other physicians have told me they prefer Sporanox for yeast-related skin conditions, and some say it is especially effective for patients with chemical sensitivity.

"In some patients, it just pulls them completely out of this complex illness that has been devastating them," says Boca Raton, Florida osteopathic physician Al Robbins, D.O. Dr. Robbins told me he uses a capsule a day for about 14–30 days for the best results, but warns that there seems to be more side effects with Sporanox than with other azoles.

In 2001, the FDA warned: Sporanox has been associated with rare cases of congestive heart failure and liver damage when used by patients with fungal infections of the fingernails and toenails. It should not be used by patients with a history of congestive heart failure or ventricular dysfunction.

Sporanox is less expensive than Diflucan, at about $3 a pill.

Nizoral (Ketoconazole)

This potent systemic anticandida drug was introduced in the United States in 1981 and was found to be effective in helping many people with yeast-related problems.

It's usually used in 200 to 400 mg. doses taken once a day.

It's by far the least expensive at about $1.25 a pill, but it does have some side effects.

The manufacturer notes: Nizoral is known to suppress hormone production, especially at high doses over an extended period of time. Nizoral has caused liver problems in one of 10,000 patients who take it, so anyone taking it should have liver function monitored monthly, more often if more than 200 mg. daily is prescribed.

If you've had a history of liver disease, I'd recommend avoiding Nizoral. It has been shown to be particularly toxic if used in combination with alcohol.

Nizoral should not be taken in combination with the antihistamine Seldane because the two drugs in combination can cause irregular heartbeats.

Another Azole—Miconazole

This generic antifungal is the main ingredient in Monistat, the anti-yeast cream sold over the counter in the United States. While the cream would not be effective against systemic candida overgrowth, miconazole gel is available under the brand name Daktarin in Europe, Australia and Canada and, when taken orally, may be helpful in treating vaginal yeast infections and thrush. Some of my colleagues say it is an excellent affordable alternative to the expensive azole drugs available in the United States.

It is relatively side-effect free compared to its azole cousins, with the only serious warning that it should not be taken in combination with blood thinners like warfarin (Coumadin).

A few physicians are familiar with this drug, but it is not easy to find in the United States. You can get it with a prescription through Wellness Health and Pharmaceuticals in Birmingham, Alabama (800) 227-2627.

More Azoles Are Coming

Within the coming year, look for new drugs in the azole class to be approved for the market. They have been developed mainly to address serious fungal infections in people whose immune systems are compromised.

In January 2002, Dr. Thomas J. Walsh told the Infectious Disease Society of America that Pfizer will soon begin marketing Vfend (voriconazole), and Schering-Plough is seeking FDA approval for posaconazole and Bristol-Myers Squibb for ravuconazole.

What's more, Dr. Walsh, chief of the immuno-compromised section of the National Cancer Institute in Bethesda, Maryland, predicted more antifungal agents called echinocandins that target the fungal cell wall, will soon be available to the public. Cancidas, the first echocandin produced by Merck and Co., recently gained approval but as of 2005, it is only available by IV administration.

OTHER EXCELLENT PRESCRIPTION ANTIFUNGALS

Nystatin (Mycostatin)

Nystatin predates the azoles as an extremely effective antifungal. Available in the United States since 1951, nystatin continues to be an effective and safe antifungal medication. However, a number of physicians and pharmacists are concerned about the tremendous increase in its price during the past two years.

Nystatin works in a unique way to help reestablish the strength of the intestinal walls, eliminating leaky gut syndrome.

The beauties of nystatin are many: It is virtually side-effect free, with the exception of occasional minor gastrointestinal upsets, and the price is by far the cheapest of the prescription antifungals, at about 60 cents for a 500,000 IU pill. Most physicians prescribe one or two capsules daily. Patients also rarely report a die-off reaction, and many say it is the only antifungal most patients need, while some others contend it is less effective than the azoles. Other physicians will start patients on the azoles and then switch them to nystatin after two or three weeks.

The downside is that some strains of candida are resistant to nystatin.

I researched nystatin in the *Physician's Desk Reference,* which contains thousands of prescription drugs, and found it was one of the safest drugs on the market.

Lamisil (Terbinafine Hydrochloride)

This drug is a member of a class of antifungals that have a different action mechanism than the azoles. It was approved in 1995 for the treatment of fungal infections of the nails. It is dispensed in 250 mg. tablets.

Lamisil has side effects similar to the azoles. Most common is a skin rash, but there have been cases of liver damage, and it has been known to affect the retina of the eye. It also has been associated with tumor growth and may interact with other prescription medications. Lamisil also can cause elevated white blood cells counts, so anyone with an impaired immune system should be monitored carefully.

During the past several years, several of my colleagues have told me that Lamisil has helped a few of their patients who had not been helped by other antifungals.

Thoughts from Other Health Care Professionals: Many physicians prefer to give patients one of the azoles plus nystatin because the combination works more quickly and efficiently to knock out the overgrowth.

Dr. Elmer Cranton of Yelm, Washington, explains why this is a more effective method of treatment:

> . . . (Diflucan, Sporanox and Nizoral) are all absorbed in the upper GI tract and taken into the circulation. They go to many parts of the body, including the liver and are excreted in the bile. Yet, the drugs are reabsorbed in the upper part of the small intestine. As a result, therapeutic levels do not reach the colon. Therefore, I strongly urge that nystatin be given in conjunction with the systemic drugs.
>
> I personally prefer to use Sporanox and nystatin together for two months. When these drugs are stopped, the benefits are lasting and my patients get well. The majority not only get well, but they stay well. They don't relapse as they often do when only one of these medicines is used at a time.

Geraldine Donaldson, M.D., in Livermore, California, has used miconaznole with great success for similar reasons:

> I also use miconazole, which I order from Canada. It costs $40 for one hundred 250 mg capsules and $5 for shipping. I've been using it for 10–12 years. Fluconazole is reabsorbed from the small intestine so that very little of it gets into the large intestine. A number of patients whose symptoms improved relapsed after fluconazole was stopped. The miconazole helps eradicate the fungi in the large bowel. I recommend starting the antifungal gradually, beginning with fluconazole and then adding miconazole because of die-off reaction. I continue both of these antifungals together until the patient recovers.

Dennis Remington, M.D., of Provo, Utah, strongly favors the azoles. He told me he usually uses Diflucan and Nizoral because stool tests he receives from the Great Smokies Diagnostic Laboratories indicate that these drugs work 500 to 1,000 times better than the nonprescription antifungals, including Tanalbit, uva ursi, caprylic acid or undecylenic acid.

He also said when he first started treating patients with yeast-related problems he used nystatin almost exclusively, but he's now finding that the stool tests show that a lot of people are resistant to nystatin. He also has found that Nizoral is more effective than Diflucan, not to mention that it is much less expensive.

MY COMMENTS

I consider the azole medications important, safe and effective for most people.

While there are side effects, taking these drugs when they are needed, based on your medical history, is much safer than letting your problems go untreated.

Yet, no drug provides a "quick fix" for candida-related health problems. You must also eat the proper diet and take other measures to regain your health and get your life back on track.

REFERENCE

1. Como, J. A., and Dismukes, W. E., Oral azole drugs as systemic antifungal therapy. *N. Engl. J. Med.*, 1994, 330:263–272.

CHAPTER 29

Nonprescription Antifungals

Nonprescription agents such as herbs, vitamins and minerals, are effective alternatives for many women with yeast-related problems who prefer not to use prescription antifungals or who cannot use them for a variety of medical reasons.

Clearly if you have maintained good health with the help of natural remedies, you're doing fine.

But people who are doing fine are not likely to be reading this book, so we'll look at the most powerful natural substances to help eradicate yeast, strengthen your immune system and help you regain your health.

Here are some of the best:

PROBIOTICS

Probiotics are a group of friendly bacteria that help us stay well. They include *Lactobacillus acidophilus* and *Bifidum bacterium*. These bacteria, which are found in natural yogurt, can help keep yeast from overgrowing in the digestive tract.

Although probiotics are not powerful agents for knocking out yeast, they do help.

Probiotics are available in a variety of forms, including yogurt and other foods, capsules, tablets, beverages and powders.

Consumerlab.com, the independent, product-testing facility, says, "Ideally the product should contain bacteria that research shows can survive passage through the stomach or it should be enteric-coated.

Products in tablets should also be able to properly disintegrate so as to release the probiotic bacteria and not pass through the body intact."

The key to using probiotics is to find the right product to help restore the natural balance. Experts tell me the ideal product will have three qualities:

1. It will have an adequate number of live organisms
2. The organisms must be bile resistant
3. The organisms must be able to adhere to the gastrointestinal lining.

You'll want to find a product that has at least 1 billion—yes, billion—bacteria per serving. There are some that claim as high a content as 60 billion organisms, but many of those organisms do not survive sitting on the shelf. Refrigerated products are better because more of the bacteria will survive.

Consumerlab tested the following products and found them to contain at least 1 billion live bacteria at the time you would consume them:

- Puritan's Pride Inspired By Nature Milk-Free Acidophilus
- Pilgrim's Pride Inspired By Nature Potent Acidophilus with Pectin
- Vitamin World Naturally Inspired Milk Free Acidophilus.

There are many good products on the market, but these are simply the ones tested and given the Consumerlab stamp of approval.

OLIVE LEAF EXTRACT

Although I've had no experience in using olive leaf extract, I've received favorable reports from people who have called and/or written me.

I was also impressed by the information in Dr. Morton Walker's 1997 book *Olive Leaf Extract* that suggests the antifungal, antiviral and antiparasitic nature of olive leaf extract can be very helpful as part of a treatment regimen for candidiasis.

He wrote:

Based on my research, I am convinced that olive leaf extract is destined to become the most useful, wide spectrum, antimicrobial herbal ingredient of the 21st century.

It is made from the leaves of olive trees and contains plant pharmaceuticals, including a phenolic compound called oleo oleuropin, a potent antimicrobial that inhibits the growth of many types of microorganisms, including viruses, bacteria, protozoa, fungi and yeast, and helps improve immune system function.

Here's a testimonial from a woman who wrote to me a few years ago:

I was troubled with multiple symptoms, some of which were back and neck pain, flu, flu-like symptoms, swollen glands, sinus and digestive problems. I was subsequently diagnosed with fibromyalgia (chronic fatigue syndrome), and the physicians recommended Prozac-type antidepressants and anti-inflammatory drugs, but I refused them.

On a treatment program that included olive leaf extract, three tablets four times a day, plus vitamin and mineral supplements, my overall health has greatly improved and so has my energy and disposition . . .

I've found some information that indicates olive leaf extract may cause die-off, so if you feel worse for a few days after you begin taking it, take comfort in knowing it's working! Even at very high dosages, there have been no reports of side effects from using olive leaf extract.

You may have to experiment to determine the correct dosage for you, but most people get good results with one to four capsules taken six hours apart, for a total of four to 12 a day.

KOLOREX

Some say this natural product from New Zealand is as effective as nystatin. It's made from anise seed and a leaf called horopito (*Pseudowintera colorata*), a relative of the cayenne pepper, that work together to restore microbial balance in the gastrointestinal tract.

Forest Herbs Research of New Zealand, the manufacturer of Kolorex, explains how it works:

> The main active component in horopito is polygodial, but at least three other natural antifungal compounds are present in the leaves.
>
> The anticandida effect of polygodial is increased 32 times by the addition of anethol, the main active of anise seed. Anise seed (aniseed) is a traditional medicinal herb in its place of origin— South America.
>
> In addition to its antifungal properties, horopito also contains the powerful antioxidant flavonoids—quercetin and taxifolin.

According to Irv Rosenberg, Ph.D., a nutritional pharmacist at the Apothecary in Bethesda, Maryland, this product is as effective as nystatin. Dr. Rosenberg sent me a copy of an article by Arnold Fox, M.D., entitled "Kolorex: The New Cure for Candida." Here are excerpts:

> From New Zealand comes a successful easy-to-use, nontoxic and natural product that effectively treats candida infections . . . Kolorex damages the cellular walls of *Candida albicans* (as well as other yeasts). With their walls ruptured, candida's cellular material leaks out and the organism dies. Kolorex kills candida more rapidly and effectively than most medicines.

Dr. Rosenberg, who has been counseling yeast patients for the past 19 years, said he believes Kolorex kills candida more rapidly and effectively than most medicines.

Dr. Rosenberg tells me he prescribes one dose a day for five days in order to lessen the amount of die-off reaction. Then, he recommends taking it twice a day for 8–12 weeks.

The company recommends taking Kolorex with food and a full glass of water. The product comes packaged with separate capsules of horopito and anise. Kolorex also comes as a cream for vaginal yeast infections. It is available through the company's website at www.kolorex.com.

CAPRYLIC ACID

This substance, a natural occurring fatty acid, was first studied some 40 years ago by Dr. Irene Neuhauser of the University of Illinois, who found that it had antifungal activity. Since it is readily absorbed, patients should take time-released or enteric-coated formulas to allow for a gradual release throughout the entire intestinal tract. Because caprylic acid is a food product, it's available without a prescription.

Caprylic acid is produced by the body in small quantities and can be extracted from plant fats, such as coconut oil and palm oil. Some research suggests it interferes with the growth and reproductive processes in the yeast organisms.

It usually requires three to four months of taking caprylic acid, 1,000 to 2,000 mg. three times a day with meals to get results. Many physicians will suggest starting at a somewhat lesser dosage, 500 mg. once or twice a day, to avoid the discomfort of massive die-off, then increasing the dosage gradually to the maximum recommended dose.

During the past decade, a number of health professionals have found that caprylic acid products are effective in controlling candida in the intestinal tract.

GARLIC

Garlic has been widely used for medicinal purposes for centuries. For example, Virgil and Hippocrates mention it as a remedy for pneumonia and snakebite. In looking through the Index Medicus, I found numerous articles from American and foreign literature describing the inhibitory action of garlic on candida organisms.

In studies carried out several years ago, Dr. Benjamin Lau of Loma Linda University, Loma Linda, California, reported on the effectiveness of Kyolic-aged garlic extract against *Candida albicans* infections in mice. According to Dr. Lau, this study suggests that the garlic extract strengthens the immune system by helping the body's white blood cells gobble up enemy germs.

As you probably know, many different garlic products, including odor-free garlic, are available from health food stores. Like the various automobile makers, each company describes the features of its garlic products. Which preparations are best? I don't know. Yet, I've been impressed by reports that document the quality of Kyolic-aged garlic extract. More than 200 peer-reviewed studies have confirmed its efficacy.

The usual dosage of garlic extracts is 400–600 mg. twice a day. You can take capsules to avoid the breath odor and/or eat fresh garlic. Of course, you can add a lot of flavor to your foods by using as much as you like.

TANALBIT

Several health professionals I've consulted rank Tanalbit high on the list of nonprescription substances they use in treating their patients.

When taken orally, this product is said to effectively destroy harmful bacteria and fungi in the digestive system without attacking friendly organisms. It also travels to the colon relatively intact to help address problems there. It seems to act as a sort of internal antiseptic. In addition to natural tannins, this product contains zinc.

Tanalbit contains tannates that bind to the surface of the fungal membranes, causing the fungus to lose its ability to stick to intestinal walls, reproduce, and then it eventually kills the fungus. Tanalbit also contains chitin, which has been shown to stop candida colonization.

The recommended dosage is two-to-three capsules three times a day with meals for three to six months for chronic candidiasis.

My long-time friend, Dr. James Brodsky, said:

I still recommend Tanalbit for patients who are not doing well with nystatin and/or Diflucan. I do not usually recommend nonprescription antifungals during the first month or two of prescription therapy. For those patients who are not doing as well as expected by the end of the second month, I recommend the addition of Tanalbit, two-to-three tablets, three times daily with meals.

CITRUS SEED OR GRAPEFRUIT SEED EXTRACT

Like nystatin and caprylic acid, this antifungal agent discourages the growth of *Candida albicans* in the intestinal tract.

Some of my colleagues also credit citrus seed extract with being as effective as nystatin in treating candida-related yeast problems. It is also effective against giardiasis and some of the other intestinal parasites.

Grapefruit seed extract (GSE) is truly a broad-spectrum natural antibiotic, capable of killing a wide variety of pathogens. It has the unique quality of being effective against yeast, bacteria, viruses and parasites. (**Warning:** Grapefruit seed extract can interfere with the effectiveness of some medications. Be sure to discuss this with your doctor.)

GSE is sold both as "Grapefruit Seed Extract" and "Citrus Seed Extract." Unless otherwise indicated on the label, these products are usually made from grapefruit seeds only.

The recommended dosage is 500 mg. three times a day.

A number of health professionals I've talked to recommend Tricycline, a product made by a company called Allergy Research, which combines grapefruit seed extract and the herbs artemisia and berberine to help achieve balance.

PARACAN

This natural product incorporates a blend of herbs that have been used for decades to eliminate microbes and worms. Its main active ingredients include black walnut hulls, wormwood, pumpkin seed, Pau d'Arco, echinacea, barberry, gentian, garlic, olive leaf, cloves, chamomile and thyme.

I've also received reports that several of the herbs contained in this product help control the overgrowth of *Candida albicans* and other yeasts in the digestive tract.

According to the web page, www.genhealth.com, "ParaCan . . . is an effective and . . . safe way to achieve optimal internal health.

"However on some occasions where parasitic activity is high, some individuals may experience nausea, weakness or fatigue during the cleansing process."

Adults and children over 12 years of age should take one capsule three times daily for one month. Thereafter repeat at least every three to four months.

Although this product is not readily available in the United States, it can be ordered from Genesis Health Marketing in Australia. Fax: (612) 9663-5310; e-mail: sales@genhealth.com.

OREGANO OIL

I first heard about the antifungal powers of this perennial pungent herb of the mint family from Ann Fisk, R.N., and Jeffrey S. Bland, Ph.D., who discussed oregano in *Preventive Medicine Update.*

These researchers concluded that oregano is an effective anti-yeast agent and more potent than caprylic acid.

A 1998 Cornell study showed that oregano was extremely effective against bacteria and fungi of all types.

I've also received favorable reports from several health professionals who are using oregano essential oil in treating patients with yeast-related disorders.

Oregano oil is sold as a liquid or in capsules. The recommended dosage is 50 mg. four times a day.

COMMENTS BY OTHERS

Michael Murray, N.D., a nutritionist and author of *Chronic Candidiasis: The Yeast Syndrome,* said he feels most comfortable in recommending caprylic acid, garlic, enteric-coated volatile oil preparations and berberine-containing plants. These plants include goldenseal, barberry, Oregon grape and golden thread. He said that berberine has been extensively studied in both experimental and clinical settings for its antibiotic activity against bacteria, protozoa and fungi, including *Candida albicans.*

Jodi Smith, an Indiana diet and nutritional consultant, told me how she helps people with yeast problems who consult her.

She emphasized the importance of changing the diet and rebuilding the immune system. She said people with yeast problems need to understand the extreme importance of the total picture rather than just taking antifungal medications. In our continuing discussion she said:

I've heard that the yeast is able to mutate and may become resistant to antifungals. So I stay away from products that contain a large variety of antifungals. I try to use one thing at a time. I start my clients out on garlic to get them acclimated to the diet and to understand what die-off is like. Then I've used oregano oil, grapefruit seed extract, olive leaf extract very successfully in some people, and of course the prescription antifungals as well.

MY COMMENTS

It's important not to attempt to self-treat these problems, even with nonprescription medications. You've probably already been through a battery of tests to rule out the easier-to-diagnose conditions such as parasites, thyroid disease and lupus. If you haven't been through these tests, do so before starting any medications.

And remember that no antifungal agent will help you if you do not change your diet and take the other measures I've described in this book.

Herbal remedies sometimes take longer to become effective. A good rule of thumb is to take an herb for at least three months before you decide whether or not it is working for you.

If your yeast problems are not responding to nonprescription agents, I suggest that you seek help from a medical doctor who can prescribe Diflucan.

CHAPTER **30**

Lifestyle Changes

Wouldn't it be wonderful if you could take a pill to erase the laundry list of problems we've discussed in this book? But I don't need to tell you there are no magic pills.

You can begin to overcome some of your symptoms when you change your diet, get rid of chemical pollutants in your home and start on antifungal medications.

> There are also many other things you'll need to do. You'll need to take charge of your health and say to yourself, "If it's going to be, it's up to me."

I think Dr. George Miller, a Pennsylvania gynecologist, passed on these thoughts about patients who complain of feeling "sick all over":

> Chances are good—even excellent—that you can regain your health and get your life back on track, and I will help you. I'd like to tell you about the 70/30 rule that I go by. This means that 70% of what needs to be done will be your responsibility and 30% will be mine.

In several of my previous books, I've casually mentioned "lifestyle changes," including exercise, stopping smoking, spending less time in front of the TV and more time outdoors.

ADDRESS YOUR FEELINGS OF ISOLATION

There are probably millions of women with candidiasis in the United States alone.

I think it is important for people with serious chronic illnesses like candidiasis to have contact with other people who understand the stresses of such life-altering health problems and to draw on the understanding of friends and family.

Feelings of isolation are not at all unusual, but isolation in itself can be very harmful to your health, because, among other things, it is stressful, and stress has been found to be an underlying cause in many disease conditions.

Dean Ornish, M.D., the well-known California holistic physician and author of *Dean Ornish's Program for Reversing Heart Disease,* suggests that people who deliberately isolate themselves from others are likely to suffer even worse health problems. Among the scientific studies he cites supporting this view is an Ohio State University study that connected loneliness with poor immune function. The *New England Journal of Medicine* reported that men who had survived heart attacks but remained socially isolated had more than four times the risk of dying of heart disease than those who have supportive networks of family and friends. Another study from Stanford University showed that women with advanced breast cancer doubled their survival rate if they attended weekly support group meetings.

Dr. Ornish says isolation can take many forms, including isolation from other people, isolation from our own thoughts and feelings, and isolation from a higher power.

"Isolation can lead to illness where intimacy can lead to healing."[1]

Decrease your sense of "apartness" by joining a support group, an exercise group, a yoga class or any course at your local college. Learn to meditate and relax.

See www.yeastconnection.com to find support, connection, and help.

In his 1998 book *Power Healing,* Dr. Leo Galland discussed "the four pillars of healing." These pillars include diet, exercise and getting rid of internal toxins that play a part in making people sick. His fourth pillar focuses on interpersonal relationships and how they play a part in enabling people to get well.

I especially like his discussion of the qualities of a caring doctor, including the ability to listen, willingness to acknowledge the patients' ideas and feelings about their illnesses, ability to show empathy and willingness to offer encouragement, hope and assurance.

PSYCHOLOGICAL SUPPORT

Seek support at this difficult time. Hopefully, you've found a kind and caring physician who is willing to listen and cares for your emotional well-being. If not, you might want to broaden your search for the right doctor.

The support of family and friends is essential when you are faced with multiple symptoms. If that support isn't as great as you would hope, you might want to give them copies of this book to help increase their understanding.

Other support sources include support groups in the larger cities, online support groups and, of course, the assistance of your spiritual leader.

Make use of them. They'll help ease your burdens.

MIND-BODY CONNECTION

Depression, stress and extreme fatigue can contribute to a negative attitude. Dozens of studies have shown that illness can have psychological causes. The illnesses are real, they're not in your head, but this underscores the power of the human mind to affect the body's wellness process.

If you're suffering from candidiasis, this is an excellent reason to deliberately encourage a positive mental attitude in the mode of Norman Cousins, editor of the *Saturday Review,* who changed himself and recovered from a life-threatening illness, at least in part through the power of laughter. Cousins' experience, documented in *Anatomy of an Illness* and *The Healing Heart* makes fascinating reading.

Recall the story I told earlier about how Norman Cousins used humor to relieve pain.

In her book, *Women's Bodies, Women's Wisdom,* Dr. Christiane Northrup offers ways for women to heal their bodies by listening to their bodies' own wisdom:

- Respect and release your emotions.
- Learn to listen to your body.
- Gather information.
- Forgive.
- Acknowledge a higher power or inner wisdom.

See **www.yeastconnection.com** for a printable list to help you notice and name your emotions.

EXERCISE

Whether your health problems are yeast-connected or not, you need to exercise if you want to enjoy good health.

Exercise helps you overcome depression and fatigue. Current scientific evidence shows exercise is a part of the treatment program for patients with depression and chronic fatigue. Aerobic exercise, including walking, running, aerobic dance, swimming and bicycling, is an excellent way to overcome the stress and the exhaustion that accompanies most yeast-related conditions.

In her book *Natural Highs,* Dr. Cass explains the science behind the benefits of exercise, from improved circulation to enhanced production of feel-good neurotransmitters including endorphins.

Exercise experts say the key to sticking with an exercise program is finding something you really enjoy. If riding your exercise bike is sheer drudgery, don't do it. Maybe you really enjoy the beauties of nature and would rather have a solitary, early-morning walk every day. If the idea of walking your treadmill alone depresses you, join a gym and laugh it up in an aerobics class.

A bonus for women: Not only does exercise help with weight control, weight-bearing exercise such as walking, running or aerobics will help strengthen your bones and prevent osteoporosis.

Population-based studies also show that women who get minimal amounts of exercise (20 minutes a day three times a week) lower their risk for heart disease and even cancer.

I personally recommend that everyone needs at least 20 minutes of exercise every day that elevates your heart rate and causes you to perspire lightly.

 See **www.yeastconnection.com** for a questionnaire to help you learn to enjoy exercise.

MEDITATION AND RELAXATION

We all live busy lives and suffering from a chronic disease makes life more difficult.

Humans were designed with a "fight or flight" response to stress. In the days of the cave dwellers, fight or flight meant making a decision to run from a saber-toothed tiger or stay and have the strength to fight it. Our bodies are also designed to quickly respond to either need, increasing heart rate, flooding muscles with blood, raising blood pressure, and temporarily taking energy from the immune system and digestive processes.

In today's world, we don't have saber-toothed tigers, but we combat the stresses of daily life as if we were fighting these invisible enemies all day long—and our bodies respond in the same way, often with toxic results.

Recent research suggests that women may have an additional hormonally driven response to stress. UCLA psychology professor Shelly Taylor, Ph.D., has coined the term "tend-and-befriend" as a more apt description of women's response to long-term stressful situations.

While women may respond to immediate dire threats with the fight-or-flight response, Dr. Taylor found that women—even female animals—under stress also nurture themselves and their young ("tending") and form alliances with others ("befriending"), perhaps by releasing a hormone called oxytocin that promotes relaxation and clear thinking at times of crisis.

If we don't learn to cope with the physiological responses our bodies make to stress, we are constantly in a state of lowered immune response with heightened demands on our hearts and glucose metabolism, leading inevitably to chronic disease.

There are many books on meditation and relaxation, and there are many methods to accomplish this. None of them are terribly complicated.

As with exercise, try several methods and choose one that you enjoy and that helps you start to feel better.

Relaxation doesn't have to be structured. Maybe the best form of relaxation for you is a long hot bubble bath at the end of a hectic day. Maybe it's simply putting on your headphones and losing yourself in a favorite piece of music or sitting quietly watching the sunset.

A wonderful form of relaxation is to progressively tighten and release your muscles from head to toe until your body feels heavy and completely at peace. Remain that way for 15 or 20 minutes. This is also a great way to fall asleep if you have insomnia.

Meditation usually involves sitting still and clearing your mind of extraneous thoughts and worries.

Here is a very basic form of meditation:

1. Choose a mental focusing device. Many people like to use a word or phrase, such as: "Hail Mary, full of grace," "Shalom" or even "One." The word you choose should be positive, but beyond that, the words are unimportant.
2. Sit quietly and comfortably. Some soft music might help you relax.
3. Close your eyes to help you focus.
4. Relax your muscles.
5. Breathe slowly and natural and mentally repeat your focus word or phrase.
6. Assume a passive attitude. Don't be discouraged if distracting thoughts come your way. Just let them float by rather than mentally commenting on them.

7. Continue doing this for 15–20 minutes once or twice a day. Make the time for yourself. Think of how much time you waste worrying, and spend these minutes constructing a positive mental attitude.

In *Natural Highs,* Dr. Hyla Cass states that, "Not only is meditation great for your state of mind, it also has many positive benefits for your body, including better responsiveness to stressful events and quicker recovery, reduced heart rate and blood pressure, a slowed rate of breathing and more stable brain-wave patterns.

Meditation has also been shown to prevent the depression of the body's immune responses that occurs with stress. People who practice meditation on a regular basis have been found to be less anxious, and there is little doubt that meditation and relaxation techniques are effective in dealing with anxiety, stress and insomnia. This confirms research at the University of Massachusetts Medical Center that found that meditators have lower levels of the stress hormone cortisol.

We have seen how even the mildest conflicts can generate an exaggerated "fight or flight" stress reaction (more appropriately reserved for dealing with life-threatening situations). Meditation can be the way to help us unlearn this conditioned response and become less reactive to the normal stresses and strains of life.

MY COMMENTS

Doing all the things you need to do to create health won't be easy, but you can get started today. Remember Confucius' proverb: "A journey of a thousand miles begins with one step."

REFERENCES

1. Ornish, D., *Dean Ornish's Program for Reversing Heart Disease,* Ballantine Books, 1990; pp. 85–103.

2. Cass, Hyla and Holford, Patrick, *Natural Highs,* Avery Penguin Putnam, 2002; pp. 211–28.

CHAPTER 31

Eating Right

You've gotten the message about eating the sugar- and yeast-free diet to address the problem of yeast overgrowth, and if you're following that diet, you're most likely eating in a healthy way.

Now I'd like to throw out a few basic ideas about healthy eating.

Over the years, Americans have become very confused about this topic. There's the high-fat low-carbohydrate diet. There's the high-fiber, low-fat diet. There's the cheeseburger diet, the grapefruit diet, the good fat versus bad fat diet. I'm sometimes confused myself. I certainly don't know all the answers, but I can give you the bare bones of what I consider some of the best approaches to healthy eating.

Shop Around the Outer Perimeter of Your Supermarket where you'll find the fresh foods. The interior aisles are crammed with processed foods. And, as I mentioned in previous chapters, rotate your foods so you eat a wide variety of fruits and vegetables.

Eat Lots of Fruits and Vegetables. Eat at least five servings a day, nine or 10 is better yet. Of course, if you're doing the anti-yeast special diet, you'll want to keep your fruit intake very low.

There are dozens of scientific studies which show that people who eat the largest numbers of fruits and vegetables are generally healthier and live much longer. Make fruits and veggies the mainstay of your eating plan.

Eat Your Fruits and Vegetables as Close to the Natural State Possible. This means fresh and raw is best, frozen is second best and canned is a far distant third (except tomatoes and tomato sauces, which actually have more healthy lycopene when they are cooked).

If you can grow your salad greens and tomatoes and other vegetables, that's even better, since they are as fresh as possible and you can assure they are pesticide-free.

If you aren't able to have a garden, buy your produce at your local farmer's market. It hasn't traveled far to get to you, which means its vitamin and mineral content will be high. Buy organic if you can to avoid the pesticides and chemicals.

If you have no other choice, look for organic fruits and vegetables at your supermarket. The problem with these is that they have often traveled a long distance to get to you, so their nutritional content has diminished. Avoid non-organic vegetables and fruits at the supermarket since they have most likely been sprayed with all sorts of pesticides and fungicides, the residue of which can remain in the food. However, eating fruits and vegetables washed carefully is much better than not eating any at all.

Eat a High-fiber Diet Rich in Whole Grains. In addition to all those fruits and vegetables, which are great sources of fiber, look for whole grains. Throw away all white-flour products, white rice and processed foods in your pantry. Not only are they detrimental to yeast elimination, they're simply not healthy.

If you're not sensitive to these grains, use whole wheat, brown rice, stone ground cornmeal, oats, barley, quinoa, amaranth and buckwheat. Become an avid label reader, and if a label says a product contains wheat flour—this is a euphemism for white flour. Don't buy it! (For more information on cooking some of the more unusual grains, see *The Yeast Connection Cookbook* by William G. Crook, M.D., and Marjorie Hurt Jones, R.N. 2001.)

Eat Modest Amounts of Good Fats. Nut and seed oils like olive oil, canola oil, safflower, walnut, sunflower, sesame and flaxseed are the fats you should seek out. They're all great sources of essential fatty acids. Be sure to get unprocessed, unrefined oils that have EFAs intact. Oils tend to become rancid when they remain on a shelf at room temperature, so purchase your oils from a store with a rapid turnover and keep them refrigerated.

Avoid Hydrogenated Oil Products like Crisco or any other spread that remains solid at room temperature. New government labeling requirements will also let you know the trans fatty acid (TFA) content of foods. Basically, you want to avoid TFAs completely.

Keep Your Meat and Dairy Consumption Fairly Low. In general, a vegetarian diet is very beneficial for almost everyone, but if you still want to eat meat, look for organic chicken and beef.

Much of the non-organically produced meat supply contains antibiotics, which can worsen your condition if you have yeast overgrowth. Insecticides are also commonly used in beef and pork. It's given to the animal before slaughter and after slaughter as a preservative.

Natural yogurt is great—the plain, unsweetened kind. I recommend low-fat varieties for general good nutrition and to help rebalance the intestinal flora.

Eat Seafood in Moderation. Seafood is good—but the pollution of the world's oceans, especially with heavy metals like mercury, has entered the food chain and made it inadvisable to eat seafood more than three times a week. Unfortunately, the best nutritional sources of seafood like salmon and tuna that have high levels of essential fatty acids are also the ones with the highest concentrations of heavy metals. Be sure to rotate the types of seafood you eat. Trim away the fat and dark meat. Also avoid stews that call for the whole fish since toxins accumulate in the internal organs. Avoid raw shellfish entirely if you have cancer, diabetes or any disease that impairs immunity.

Avoid Tap Water. Tap water often contains chemicals from a variety of sources. These include insecticides and weedkillers that remain in the water in spite of purification and filtration. Fluorine, chlorine and other chemicals are added to the water, and chemicals are picked up from plastic or copper pipes. Use well, spring or distilled water.

Don't Microwave Food in Plastic Containers. Plastic can actually release toxic substances into the food, especially into fatty foods.

MY COMMENTS

Again, this is only the bare bones primer of healthy eating. This is what everyone should do, regardless of your health condition. If you have yeast-related problems, you'll need to be much more vigilant about all the types of food you eat, at least until you've controlled the yeast overgrowth and identified your hidden food sensitivities.

RESOURCE

Crook, William G., and Jones, Marjorie Hurt, *The Yeast Connection Cookbook.* Professional Books, 2001.

CHAPTER **32**

Nutritional Supplements

I f you have yeast-related health problems, there are some specific supplements that may help you return to good health, and there are others that simply make good nutritional sense. Regaining your health depends not only on the health measures you take specifically to target the yeast, but on those to help your general health and particularly to strengthen your immune system.

If you have yeast-related health problems, you'll want to be sure to buy yeast-free, sugar-free, color-free multivitamin, minerals and anti-oxidant preparations.

 See **www.yeastconnection.com** for a convenient source for these products.

MULTIVITAMINS

In a dramatic about-face, the American medical establishment now says that all adults should take a daily multivitamin. Two decades ago, that wasn't the case. The medical authorities actually advised against taking multivitamins after studies concluded there was no evidence to suggest any significant benefits to people who took them.

But in June 2002, the medical establishment did a 180-degree turn with the publication of a Harvard study recommending that all adults take a multivitamin every day. The authors of the study, published in the prestigious *Journal of the American Medical Association* concluded that, while few Americans are vitamin deficient, only 20 to 30% of the population eats the recommended five servings of fruits and vegetables a day, so multivitamins can help prevent chronic diseases. They recommended that all adults take a daily multivitamin to help prevent heart disease, cancer and osteoporosis. In addition, women in their childbearing years should consider adding folate, and the elderly might add vitamins B12 and D to their daily multivitamin.

I couldn't agree more wholeheartedly. However, you probably know my position well enough by now to expect what I'm going to say next: Taking a bunch of vitamins pills and continuing to eat diets loaded with junk foods does not promote good health.

I take vitamin and mineral supplements and recommend them to the adult members of my family.

When supplements are prescribed by a knowledgeable professional, the amounts may vary considerably from those I've outlined on page 240. Your physician's expertise, experience and clinical judgment should override my recommendations.

ESSENTIAL FATTY ACIDS—THE GOOD FATS

Our bodies are composed of billions of cells of various sizes, shapes and functions. Each cell is surrounded by a membrane composed of special types of fats called essential fatty acids (EFAs).

These good fats come directly and only from our foods and have many functions. Moreover, they are important in preventing health problems of many types, including heart disease, arthritis, PMS, eczema and other skin disorders.

There are two general classes of EFAs:

1. The Omega-3 fatty acids like those found in cold-water fish like salmon and tuna, fish oil and flaxseed oil.
2. The Omega-6 fatty acids found in meats, safflower, canola and nut and seed oils.

For my adult patients with yeast-related health problems, I recommend these daily supplements at a minimum. These nutritional supplements strengthen your immune system. You should begin taking them at once:

Vitamin A	5,000 IU
Beta carotene	5,000–10,000 IU
Vitamin B1	25–100 mgs.
Vitamin B2	10 mgs.
Niacin	50 mgs.
Niacinamide	100–150 mgs.
Pantothenic acid	100 mgs.
Vitamin B6	25–100 mgs.
Folic acid	200–800 mcgs.
Vitamin B12	100–2,000 mcgs.
Biotin	300 mcgs.
Choline (Bitartrate)	100 mgs.
Vitamin C	1,000 mgs.
Vitamin D	100–400 IU
Vitamin E	400–600 IU
Calcium	500 mgs.
Magnesium	500 mgs.
Inositol	100 mgs.
Citrus bioflavonoids	100 mgs.
PABA	50 mgs.
Zinc	15–30 mgs.
Copper	1–2 mgs.
Manganese	1–2 mgs.
Selenium	100–200 mcgs.
Chromium	200 mcgs.
Molybdenum	100 mcgs.
Vanadium	25 mcgs.
Boron	1 mg.

The majority of people get plenty of the Omega-6 fatty acids but not enough of the Omega-3s. You can take supplements containing the good oils with the proper ratios for maximum health benefits. Look for a quality brand.

ANTIOXIDANTS

Cells within our bodies form free radicals. Free radicals contribute to a number of health conditions, notably heart disease and cancer. Ordinarily, the protective actions of antioxidant molecules control the negative effects of free radicals. Vitamins A, C, and E, as well as other nutrients such as selenium, magnesium, beta carotene, flavonoids, glutathione, alpha lipoic acid and many more can neutralize the harmful effects of the free radical molecules and protect our health.

Think of antioxidants as scrubbers that keep your body free of harmful oxygen free radicals and keep your cells young and healthy.

You'll find some of them in your regular multivitamin and, of course, you find them in fruits and vegetables. You may want to include some others in addition to your multi. Be sure to include antioxidants in your health regimen every day.

OTHER SUPPLEMENTS

Many other nutritional supplements may help, including:

- CoEnzyme Q10: A powerful immune system strengthener and an essential element in promoting the production of energy (adenosine triphosphate or ATP) in the body.
- Ginkgo biloba: Best known for helping maintain brain function, ginkgo has also been shown to increase energy and help combat fatigue.
- Echinacea: Another powerful immune system strengthener that is credited with making the immune system cells more efficient in combating infection. There is some evidence suggesting echinacea may help make the body more resistant to candida yeast overgrowth.

- Bromelain: Best known for helping digestion and relieving inflammation, bromelain also has been shown to be effective in relieving sinusitis and fighting urinary tract infections.
- Grapeseed extract: A strong antioxidant that helps boost immune function and protect you against a wide variety of diseases.

RESOURCES

The American Botanical Council has basic information on its website and publishes a magazine (*HerbalGram*) and authoritative books on the medicinal qualities of herbs. www.herbalgram.org.

Murray, Michael, *Encyclopedia of Nutritional Supplements*. Prima Publishing, 1996.

Hobbs, Christopher and Haas, Elson, *Vitamins for Dummies*. John Wiley and Sons, 1999.

Carper, Jean, *Stop Aging Now* and *Miracle Cures*. Perennial, 1996 and Harper Collins, 1997.

Cass, Hyla, *User's Guide to Herbs*. Basic Health Publications, 2003.

PART 6

The Yeast Connection and Your Weight

CHAPTER 33

Understanding the Weight Connection

Editor's note: *In the three years since Dr. Crook's death, his research and treatment of yeast overgrowth have received more attention and positive feedback than during his entire medical career. Now, as other doctors build on his findings, Dr. Crook's life's work is gathering momentum. Among the doctors furthering his valuable work is Carolyn Dean, M.D., N.D., health advisor to YeastConnection.com and a staunch believer in Dr. Crook's work and the condition she has named "Crook's Candidiasis." This chapter was written by Dr. Dean for this updated second edition.*

A DIET THAT WORKS

Have you tried low-fat, low-carb and every imaginable diet, yet gotten nowhere? Do you crave sugar, carbohydrates and alcohol? Do you have belly fat that you just can't get rid of?

Perhaps you picked up this book because you have other health problems. Weight may not even be your primary concern. However, excess weight *is* a concern for 65% of the American population, so the chances are that being overweight also affects your health picture. Statistically, and based on decades of working with female patients of all ages and backgrounds, I can say with conviction that there is a strong likelihood that if you do have a weight problem, it is most likely caused by yeast.

This is the same yeast overgrowth that might be behind all the other conditions you've read about in this book, including extreme fatigue, headaches, mood swings, bladder infections and poor concentration.

The reality is that yeast overgrowth, by itself, can make you fat.

The good news is that once you understand that yeast may be the culprit, following our new 6-Point Yeast-Fighting Program will help you change the outcome.

 See our website, **www.yeastconnection.com** for a detailed explanation of the 6-Point Yeast-Fighting Program and the experiences of women who have followed it.

THE SCIENCE CONNECTING YEAST AND WEIGHT GAIN

Think of the following factors that may be affecting your health and that of your children:

- Children are constantly exposed to cold and flu germs at daycare and at school.
- Doctors often prescribe antibiotics, even though these are useless against the viruses that cause colds and flu. Though antibiotics do not work preventively, many doctors still prescribe them in a vain attempt to "prevent" sinus infections, strep throat, or ear infections. They sometimes give in to pressure from distraught parents who want to give some medications to their sick children.
- As a result of taking these antibiotics, the intestinal flora become unbalanced. The good bacteria are killed off, but the yeast are not harmed.
- Yeast grows out of control in the digestive tract and toxins begin to build.
- Leaky gut syndrome may then occur, spreading yeast toxins throughout the body. (See Chapter 3 for details on leaky gut.)
- This triggers food sensitivities and intolerances, among other things, causing the vicious cycle you've already read about in Chapter 4.

- By the time a child reaches adulthood, he or she has probably experienced dozens of rounds of antibiotics and most likely is already incubating horrendous yeast overgrowth.

That's the first part of the cycle. Now let's look at how this relates to weight gain.

A Vicious Cycle

Our hypothetical young woman, whom we'll call "Mary," has reached her early 20s. She's become sexually active and has already suffered through a few bouts of vaginitis, for which she was prescribed antibiotics.

During her teens, Mary took tetracycline for a year to help keep her skin clear. You're getting the idea: Antibiotics are killing the beneficial bacteria in Mary's digestive tract and her yeast population has now begun to blossom out of control.

Add to that the toxic world Mary lives in:

- She's getting even more antibiotics through her food, since farmers feed antibiotics to livestock and the drugs are concentrated in the meat and poultry she eats.
- The same is true for the growth hormones added to livestock feed, which affect our hormones to such a degree that the average age of first menstruation for American girls is now just under 12 years— as compared to age 16 a century ago! In addition, American girls are showing the first signs of puberty at much earlier ages and experiencing early puberty-related weight gain. Just the fact that an 8-year old girl weighs 100 pounds can by itself trigger the early onset of menstruation.
- Herbicides, pesticides and fungicides are used to boost crop production and increasing amounts of these toxic substances are now present in our food chain. Some of these strong chemicals act as hormone disruptors and as cancer-causing agents.
- Toxic chemicals are everywhere, from our carpets to our cleaning supplies to our beauty products.

This isn't a pretty picture, but I'm afraid this sordid situation gets worse.

Even More Toxins

We absorb toxins into our bodies in a number of ways. We eat:

- Synthetic food
- Junk food
- Highly processed refined foods
- Refined sugar
- Additives
- Preservatives

Our bodies do not recognize these substances as food. We most often respond to chemicals in food as foreign invaders and this stimulates an allergic response. Multiply that by a dozen chemicals in just one packaged food and your body is under attack.

We now have an epidemic of autoimmune disease where the body attacks normal tissue (i.e., in multiple sclerosis, the nerves come under attack and in rheumatoid arthritis, it is the joints). When I attended medical school, there were half a dozen autoimmune diseases. Now we have identified 80.

Our defense systems aren't attacking a normal self, they are attacking a *toxic* self, where external environmental toxins and yeast toxins "gang up" on the body to create autoimmune disease.

LEAKY GUT SYNDROME

You already know about leaky gut syndrome from Chapter 4, but let me explain how this syndrome relates to weight gain.

Leaky gut syndrome takes place when the *Candida albicans* yeast in the digestive tract overwhelms the protective *healthy* bacteria by burrowing into the now-unguarded intestinal wall. This in turn creates tiny gaps in the membrane lining, through which partially digested food particles and the 180 toxic byproducts of yeast may be absorbed into your bloodstream and thus reach every part of your body.

Leaky gut can cause all kinds of seemingly unrelated illnesses, from bladder infections to psoriasis to depression and more. These toxins then create an allergic response that stimulates:

- food cravings (especially for carbohydrates, yeast's favorite food)
- bloating
- weight gain
- liver overload as the main organ of detoxification tries to deal with all these toxins
- decreased thyroid function and metabolism, since the liver is busy elsewhere
- tremendous fluid retention as the body tries to dilute the toxins
- fat cells swelling as they trap toxins to try to protect the rest of the body
- gas buildup from yeast
- toxins that impair normal thyroid hormone function
- hormonal imbalance on all levels caused by pseudohormones (false hormones) released when environmental toxins interfere with hormone receptor sites

The stress your body experiences in coping with toxins, weight gain and general physical discomfort creates excessive release of cortisol, the chronic stress hormone, making it nearly impossible for you to lose weight. The bottom line of all this is weight gain that keeps you heavy, sluggish and miserable.

THE VICIOUS CYCLE BECOMES AN ADDICTION

Yeast itself feeds on sugar, triggering sugar and carbohydrate cravings *on top of* the cravings that come from the allergic response due to leaky gut. It's almost as if yeast has a mind of its own, persuading you to feed it and help it grow, thus making you fatter and fatter and sicker and sicker.

It's not that you are weak-willed. These are genuine cravings that, like an addiction to drugs, are very difficult to overcome. In her book *No More Heartburn,* Dr. Sherry Rogers points out that people are often addicted to foods that make them fat. As she explains:

"Some folks actually have hidden addictions to foods like milk, wheat, sugar, alcohol, or coffee that drive them to stay constantly fueled with that food. Because there are actual opioid receptors (the same ones that create addictions to drugs like opium) in the brain and in the gut that make folks crave a food, they will go out of their way to make sure they have a constant infusion of it. Likewise, withdrawal symptoms, such as headache, agitation or nausea, make avoidance difficult for the few days that they last."

When you go on a weight loss diet and fat cells begin to break down, you experience another aspect of withdrawal. In the murky depths of those cells lie toxins and chemicals that have been hidden away to protect the brain and delicate organs. Once these are let loose into the blood stream, they cause havoc, leading to headaches, nausea and joint pain that can make you run screaming from any further thought of dieting.

HORMONE IMBALANCE AND IMPAIRED METABOLISM

Toxins from yeast overgrowth create allergic responses, and trigger hormone imbalance. Some of these toxic byproducts, or pseudohormones, can over-stimulate or "jam" hormone receptor sites, causing weight gain and mood swings that for many people lead to binge eating.

When your hormones are out of control:

- Your metabolism is impaired, slowing the calorie-burning process, adding weight and making it more difficult to lose weight;
- Your immune system suffers as more toxins begin to circulate in your system, increasing the toxic overload and the allergic response; and
- Your blood sugar control is impaired and high blood sugar feeds the yeast even more.

It's a vicious cycle with layer upon layer of nasty effects on your body. However, and I can't emphasize this enough: YOU CAN BREAK THE CYCLE!

That is why you are reading this book, and that is exactly what I want to help you begin to do, right here, right now.

Current research focuses on the drug treatment of medical conditions and not the lifestyle changes that are necessary to treat most health problems, most particularly yeast overgrowth.

In my practice, I treated about two thousand patients with a yeast-fighting program very similar to the one outlined in this book and at www.YeastConnection.com. The majority of people on this program experienced breakthrough results on their path to permanent well-being.

ANTIBIOTICS IN OUR FOOD SUPPLY

You're probably convinced now that raging yeast is making you fat. But there is one more source of yeast-connected weight: the antibiotics in our food supply.

Farmers add antibiotics to livestock feed to encourage the animals to gain weight and to hasten their journey to the slaughterhouse. The Centers for Disease Control and Prevention reports that more than half of all the antibiotics sold in the U.S. are used in livestock production and 90% of those are solely for the purpose of encouraging weight gain in animals.

We know these antibiotics are in our food chain because of the increasing numbers of antibiotic-resistant bacterial strains. These bacteria have made it necessary for even stronger dosages of new antibiotics to counter bacterial infections that were easily treatable a couple of decades ago.

It's not a great leap of logic to understand that these antibiotics in our own systems are not only affecting the healthy microbial balance in our digestive tracts, leading to yeast overgrowth, but more importantly, *these antibiotics that cause livestock to gain weight are likewise causing humans to gain weight!*

DR. CROOK'S OPINION

Dr. Crook firmly believed there was a causal connection between yeast overgrowth and overweight. Since he developed his theories on yeast overgrowth, the problems discussed above have only gotten worse. Certainly we have created a more toxic world.

In his books and papers, Dr. Crook often cites the discovery by Dr. Orian Truss that individuals with yeast overgrowth develop significant metabolic and biochemical abnormalities. Dr. Truss and other researchers noted that sugar cravings, fatigue and other manifestations of hormonal dysfunction are almost universal among people with yeast-related problems.

Dr. Crook also theorized that toxins entering the bloodstream as a result of leaky gut syndrome affect the pancreas, the organ responsible for producing insulin. When you drink a can of soda that contains ten teaspoons of sugar or eat sugary foods, your pancreas is stimulated to produce excess insulin and your blood sugar decreases, causing you to crave sugar—another part of the vicious cycle.

LOOKING AT THE ROOTS

In summation, think back to the beginning of this chapter and our discussion of "Mary's" lifelong challenges with yeast overload. Add to this the fact that many young women with yeast overgrowth transfer the infection to their newborns as these babies pass through the birth canal, leading to infections from birth. A series of ear infections, extremely common in babies and usually treated with antibiotics, can actually speed up the debilitating process of dysbiosis.

Now the epidemic of excess weight begins to make sense. The obesity problem may not be what it seems. Perhaps what we have is, at least in some significant measure, a national epidemic of "Crook's Candidiasis." That's certainly something to consider!

CHAPTER **34**

Get the Weight Off— and Keep It Off

Over the years, as I treated thousands of patients with yeast problems, I also observed that most of them had weight problems. For many of these patients, their excess weight was the least of their worries, since they suffered from chronic fatigue, persistent headaches, endometriosis, brain fog and a host of other serious health problems.

Yet almost without exception, those who needed to lose weight found that the weight melted off with the Yeast-Fighting Program.

These days, I am hearing from women with diet-resistant excess weight who ask for my recommendations for weight loss strategies that work. Many of these women—perhaps you are among them—do not have overwhelming symptoms of yeast overgrowth. They've just found it nearly impossible to lose weight or keep it off. I believe many of these women may, in fact, have Crook's Candidiasis.

There's a simple way to find out. You may have skipped through the early chapters of this book, eager to get to this chapter. If so, take a moment, go back to Chapter 2 and complete the candida questionnaire. When you do, if your score is over 60, consider that candida may be contributing to your weight problems. If your score is over 120, candida is almost certainly causing your weight gain along with other problems.

If you have already tried dozens of diets, met with limited success, gaining back all you lost and perhaps putting on even more weight, I advise you to take heart and follow the Yeast-Fighting Program. I suspect it may be exactly what you need to leave your tyrannical and stubbornly persistent weight-related problems behind you once and for all.

THE 6-POINT YEAST-FIGHTING PROGRAM

Since the publication of the first edition of this book we've further refined the 6-Point Yeast-Fighting Program. The plan is constantly evolving, so please regularly check our website—www.YeastConnection. com—for a wealth of continuously updated information regarding each aspect of the plan and other tested and proven success strategies.

A healthy diet is still the first step and the centerpoint of the program. As Dr. Crook always maintained, no amount of treatment for yeast overgrowth will be successful without a vastly improved diet. Here is our 6-Point Yeast-Fighting Program:

1. The Yeast-Fighting Diet

I can hear you groaning, "Oh, no, not another complicated diet that will make my life even more difficult!" The good news is, this diet program is *not* complicated.

In fact, it's quite simple:

- You can eat most fresh vegetables, nuts, lean meats, poultry and fish to your heart's content, along with a prescribed list of grains and lots of pure water;
- You will be eliminating the processed and refined foods and sugars that contributed to your illnesses and weight gain in the first place; and
- You'll also be eliminating fruits, fermented products and foods containing yeast, at least in the early weeks of the plan, until you have determined which foods trigger your symptoms.

 The Yeast-Fighting Program is constantly evolving, so check our website, **http://www.yeastconnection.com/** for the latest information.

The Yeast-Fighting Diet is not a starvation diet. It is somewhat low in carbohydrates, but does not restrict calories. It basically starves yeast, not you! And it works.

Let's take stock here:

- How many diets have you tried that advise you, as we do, to eat oatmeal with butter for breakfast?
- How many encourage you to eat nuts and lots of healthy oils?
- How many encourage you to eat brown rice and other high-carb grains?

It's not just by eating fewer calories or fewer carbs that you will lose weight on the Yeast Fighting Diet. It's because you're starving the yeast. ENOUGH WITH THE VICIOUS CYCLE!

The basics of the Yeast-Fighting Diet are outlined in Chapter 27. If you whizzed by that chapter, go back and read it carefully.

2. Take Nutritional Supplements on an Ongoing Basis. These Should Include:

- **Probiotics** because they promote a natural, healthy balance of microorganisms in your digestive tract.
- **Digestive enzymes** because they promote optimal digestion and ease digestive upsets.
- **Herbs and nutrients** because they inhibit the growth of candida yeast. You'll find detailed information about these in Chapter 29. A simple solution is to try the Yeast Connection Physician's Candida Formula, available through our website. This is a specially blended combination of caprylic acid, pau d'Arco, oregano oil, black walnut, grapefruit seed extract, garlic, beta carotene and biotin.
- **A good-quality multi-vitamin and mineral supplement** because they help supply your body with all the nutrients necessary to help you regain your health.

 You'll find the best quality yeast-fighting supplements at our website, http://www.yeastconnection.com/.

3. Minimize Chemical Exposures in Your Home and Workplace

Review your environment at home and at work to determine if any of the most common chemical or mold exposures might be causing difficulties for you. You'll find a detailed list of these toxic products in Chapter 19, and on our website.

Take these actions to further reduce your exposure:

- Breathe outdoor air every day and open your house and office to air them out.
- Remove as many chemicals as possible from your home.
- Use baking soda, unscented soap, vinegar, lemon juice, or food-grade hydrogen peroxide for household cleaning.
- Use unscented cosmetics, deodorants, soaps and laundry detergents.
- Dry clean as few items of clothing as possible.
- Buy and eat organic foods and products as much as possible.
- Take saunas. Sweating helps your body clear itself of toxins.

> See our website, **http://www.yeastconnection.com/** for an extensive interview with fitness expert Teresa Tapp for an exercise program that is specially designed for people with health challenges like yours.

4. Move!

How many times have you been told that the key to all your problems, especially your weight problem, is to eat right and exercise? It's true—and it's also not surprising if you are resistant to this simplistic approach.

- Begin with something as simple as spending time outdoors. Move fresh air in and out of your lungs. Transfer your attention from a TV or computer screen to a bird, a cloud, or falling rain. For some, that will be a huge beginning.

- If you can, walk or swim, do stretches and simple resistance exercises.
- Listen to your own body. Experiment. See what feels good to you because that is what you will enjoy doing.
- Check our website for advice from exercise experts Teresa Tapp and Dr. Irv Rubenstein.

 If you're having trouble finding the right healthcare professional, check out the resources section of our website, **http://www.yeastconnection. com/**.

5. Take Care of Yourself

Your body and your emotions are intertwined. Each affects the other in powerful ways. Make time for yourself and take time to care for yourself emotionally:

- Increase your awareness of how you are resisting change and understand that change is normal for everyone, and so is resistance to change.
- Then work through your resistance and "fake it till you make it." Act and the enthusiasm will follow!
- Find excellent advice and a strong ongoing support group on our website.

6. Work with a Healthcare Professional

- It helps if you can find a physician experienced in treating yeast-related problems.
- If your doctor is not experienced, but is open to new ideas, you can explore your options together by using the physician packet available through our website.
- If you do not have a physician, look for resources in Chapter 9 and on our website.
- Discuss with your doctor whether or not you need to take prescription antifungals.

OOPS! I FELL OFF THE PROGRAM!

You and I are human. None of us is perfect. It's quite possible you'll fall off the program. You may slip at a time when you're having thoughts that run along these lines:

- "This is a lot of trouble."
- "I don't think it's worth it."
- "I don't feel that much better."
- "I don't have the energy to do this."

These self-sabotaging thought patterns may lead you to indulge in some of your old eating habits. If so, you will most likely notice the quick return of some of your former symptoms. If that isn't a wake-up call, I don't know what is!

If this happens, it is a signal for you to pick yourself up and get back with the program. Don't beat yourself up, because that only reinforces the negative self-talk. Consider that you've learned a valuable lesson.

CHAPTER **35**

An Inspiring Success Story

All this talk about weight loss is intellectual until you see it applied to someone who is a real example and an inspiration. Let me introduce you to Michelle, a 37-year old woman who works as a caregiver at a training school for the developmentally disabled. She is also a budding artist. Michelle, who successfully utilized the Yeast-Fighting Program, generously, openly and honestly shares her dramatic story below.

MICHELLE'S STORY

I've spent most of my adult life profoundly depressed. I can trace it back now to the time I got my first urinary tract infection (UTI) right after I graduated from high school. Of course, my doctor put me on antibiotics. And I got more of the same every time I got another UTI—which has been about three times a year for nearly 20 years now. The doctors decided I had a short urethra and recommended surgery, but I resisted.

When I was in college, I was very depressed and unmotivated, despite my passion for my major, art. I just couldn't seem to concentrate and seemed to be walking around in a fog all the time. I couldn't keep up my studies, so I dropped out after two years.

My Downhill Slide

I don't know for sure what was wrong with me then, but now I suspect that's when the yeast problems first manifested themselves. All I know is that I felt like crap.

After I left college, I started dating a guy who was an alcoholic and verbally abusive. I think he reflected how I felt about myself then. It took me seven years to extricate myself from that toxic relationship. I was in therapy and was finally put on Prozac. I thought that would help, but it didn't really help much. It just cleared my mind a little bit. And I kept right on getting UTIs three times a year, almost like clockwork.

When I was 28, I married a great guy who was stable. For the first time in years, I felt secure. But my depression continued. We argued all the time because of my depression and mood swings. We talked about having kids. I really wanted a baby, but my husband insisted on waiting because of the arguments, which I'll admit were triggered by my irritability and depression. I agreed to put off having kids, thinking he was probably right about my inability to be a mom.

I Was a Carbohydrate Addict

During that time, we drank a lot of beer and ate a lot of pizza. I loved condiments and ate huge quantities of ketchup, mustard and hot sauce. Now I know I was just feeding my yeast with all those foods high in yeast, cheese, and sugar. I was making myself worse, but I had no idea what was going on. I accepted that even though there seemed to be something seriously wrong with me, I just had to live with it.

The arguments got to be too much and we were divorced after five years. During that time I steadily gained weight until I reached about 155 pounds. That made me even more depressed. It seemed that no matter what I did, I couldn't lose that weight.

I had picked up an herbal remedy book about weight loss that recommended allergy testing and even mentioned yeast as a possible allergy. I went to an allergist and he was dumbfounded. He had never heard of a systemic yeast allergy, but he did a whole battery of tests anyway. The test showed that I was allergic to molds and pollen. I was dying to find out about my diet and get cracking, but it took five rounds of tests before we got to food allergies.

A Diagnosis at Last!

And yes, I was allergic to yeast. The doctor told me to avoid bread and baker's yeast. So I went home, popped open a can of beer and toasted my relief that I wasn't allergic to brewer's yeast. I didn't have much trouble giving up bread, but I wasn't getting any better.

On the next visit, the allergist told me, "Yeast is yeast." I hadn't heard him correctly. I was to stop brewer's yeast as well, so I said good-bye to beer. In addition, he took me off all forms of alcohol, vinegar and bread. But I must admit, I didn't follow the diet properly.

With the guidance of a naturopath and bottles of supplements, I was able to get off the Prozac, but still didn't feel that much better.

I decided the depression was simply my personality type and there wasn't anything I could do about it. My main goal at that time wasn't to let go of the depression: It was to lose weight.

A Fortuitous Visit to the Bookstore

Everything changed the day I went to a bookstore and found *The Yeast Connection Handbook* by Dr. William Crook. (NOTE: This book is available at www.YeastConnection.com). The moment I opened the book, I recognized myself: The depression, mood swings, irritability, severe PMS, crying for no reason, needing a nap in the middle of the day.

I finally got serious about the Yeast-Fighting Program during the first week in March and I started visiting the www.YeastConnection.com website, where I found lots more information and the support of people who are going through similar health crises.

Within days of really eliminating yeast, I felt like a different person! I started losing weight at a very good pace. In just two months, the change in my mood was like night and day. I didn't feel depressed at all. For nearly 20 years, depression was like a black cloud in my head all the time. It just went away. It's not there anymore!

The Weight Is Melting Off

Not only did the depression go away, but my extra weight also started to disappear almost magically. I lost 15 pounds in about six weeks. I had tried every diet. You name it, I did it. Nothing worked until the Yeast-Fighting Program, which has given me the fastest and best results of any diet I've tried. Of course, I actually changed my whole lifestyle.

On the whole, I haven't felt this good in 20 years. I did all this without a doctor, but I would say if you are on medications don't come off them without consulting your doctor first. I followed the diet and I took some supplements that certainly helped, including a probiotic and an antifungal that included caprylic acid.

Today I wore a pair of jeans somebody gave me five years ago. At one time, I couldn't even get them over my calves. When I wore them to work, everybody noticed how much weight I had lost. Good-bye old baggy pants!

I work with about 30 women, many of whom will jump on the bandwagon with any new diet. They were amazed at my success. I can actually tell who's got a yeast problem by just watching the way they're eating—bread, cakes, etc.

My journey hasn't been effortless and it is not completely over. For example, I thought I was doing fine by eating wraps, using flatbread I had found at my supermarket. I figured since the bread was flat, it didn't have yeast. Wrong!

DR. DEAN COMMENTS

Michelle fell into the trap that many people with yeast have discovered: Food that seemed OK turned out to be not OK. Yeast, sugar and fermented products may be hidden in a wide variety of foods that you'd never suspect. Not only do you need to read labels carefully and learn to decipher the mysterious codes, but realize that some products don't even mention the added sugar. This is especially common in canned goods. There are other Yeast-Fighting Program no-nos that can sabotage your get-healthy plan, even in small amounts, so read this book carefully and keep visiting our website.

MICHELLE BECOMES A LABEL READER

Here is a bit more from Michelle on this important topic:

One day, after I was fairly well into my detox, I just smelled the yeast in the flatbread. I guess my nose had become more sensitive, so I looked at the label and, sure enough, it had yeast. I had been eating veggie burgers and thought they were fine until one day I smelled the yeast in them. Sure enough, there it was again, so I threw those away.

Now I've learned I have to be a really careful label reader. I eat a lot of chicken, turkey and tuna. I found some rice crackers I really like—and mochi (a pounded rice that you toast in a toaster oven) and spelt sticks. I've increased my vegetable intake dramatically. Plus, I drink a lot more water. I had been drinking five cups of green tea a day, but I had no idea it was just a little bag of mold. When I figured that out, I gave away all my tea!

HUGE CHANGES

My life has changed in so many ways in these few short months. I'm going back to college—majoring in art, my true passion. Being free from fatigue and lack of motivation, I'm once again using my creative skills.

Also, being less moody, depressed, and irritable, I'm more outgoing, and I no longer isolate myself, thinking, "Who the heck wants to be around me? I'm so depressed." My social life has improved tenfold. That anxious feeling I used to get prior to going to any function has also vanished.

My communication and concentration skills have improved so I no longer fear that I'll have nothing to say or that I won't be witty enough. This yeast-fighting program has helped me to enrich my life. For the first time, I am not jealous of other people, envying their social ease, wittiness and grace. Now I have that myself.

DR. DEAN COMMENTS

Michelle's story is very inspiring! She's far from the only one who has lost weight and kept it off for years with the Yeast-Fighting Program.

As a doctor, and from a personal perspective, I can see that people are more confused than ever about diet. I know it firsthand from my practice and from our message board. People are fed up with all the conflicting advice, and sadly, many are just giving up.

Too few women have even an inkling that their symptoms could be caused by yeast. We hope this book will give you a jump start to the vastly improved health and well-being that you're looking for. If it does, we hope you'll spread the word!

Concluding Comments

Throughout this book, I've offered you my thoughts and those of a multitude of other health care professionals. I hope to have helped you find a way back to health or, at the very least, to have inspired some thinking and new avenues of investigation as to the source of your health problems. I leave you with my best wishes and these final thoughts:

1. If you're fortunate, a kind, caring, knowledgeable, interested professional is working with you. Be guided by his/her recommendations.
2. It's important not to attempt to self-treat these problems, even with nonprescription medications. You've probably already been through a battery of tests to rule out the easier-to-diagnose conditions such as parasites, thyroid disease and lupus. If you haven't been through these tests, do so before starting any medications.
3. If you've experienced adverse reactions to any antifungal agent, check with your health professionals. These symptoms may be temporary (a "die-off" reaction).
4. No antifungal agent will help you if you do not change your diet and take other measures I've described in this book.
5. If an antifungal agent doesn't seem to be helping after three weeks, a different one may be more effective.
6. If your yeast problems are not responding to nonprescription agents, I suggest that you seek help from a medical doctor who can prescribe Diflucan.
7. If you're carefully following the recommendations in this book and you aren't improving, check with your physician or other

health professional to make sure that other causes of your symptoms haven't been overlooked.

Here's to your health!

William G. "Billy" Crook, M.D.

William G. Crook

William G. Crook, M.D., received his medical education and training at the University of Virginia, the Pennsylvania Hospital, Vanderbilt and Johns Hopkins. He was a fellow of the American Academy of Pediatrics, the American College of Allergy and Immunology and the American Academy of Environmental Medicine. He was a member of the American Medical Association, the American Academy of Allergy and Immunology, Alpha Omega Alpha and other medical organizations. Dr. Crook was the author of 13 previous books and numerous articles in the medical and lay literature. For fifteen years, he wrote a nationally syndicated health column, "Child Care" (General Features and the Los Angeles Times Syndicates).

Many of his publications have been translated into French, German, Japanese and Norwegian. *The Yeast Connection and Women's Health* is the eighth in his series of books which deal with the relationship of *Candida albicans* to many puzzling health disorders. The titles include *The Yeast Connection, The Yeast Connection Cookbook, The Yeast Connection Handbook, Chronic Fatigue Syndrome and the Yeast Connection, The Yeast Connection and the Woman, Tired—So Tired! And the yeast connection* and *Yeast Connection Success Stories.*

Dr. Crook was a popular guest on local, regional, national and international television and radio programs, including Oprah Winfrey, Sally

Jessy Raphael, Regis Philbin, The 700 Club, Good Morning Australia, TV Ontario and the BBC.

He addressed professional and lay groups in 39 states, all Canadian provinces, Australia, England, Italy, Mexico, The Netherlands, New Zealand and Venezuela.

He served as a visiting professor at Ohio State University, and the Universities of California (San Francisco) and Saskatchewan.

Dr. Crook lived in Jackson, Tennessee, with his wife, Betsy, where he practiced medicine for almost 40 years. They had three daughters and four grandchildren.

He compiled most of the manuscript for this book before his passing in October 2002.

Carolyn Dean

Dr. Carolyn Dean is a rarity in medicine. She is both a medical and naturopathic doctor and is one of the few people in the world to hold those titles simultaneously. She graduated from Dalhousie Medical School in Nova Scotia in 1978. She is also a graduate of the Ontario Naturopathic College, and in 2005 completed her six-year appointment as a board member of the Canadian College of Naturopathic Medicine in Toronto.

Dr. Dean began practicing integrative medicine in 1979, long before that term was even invented. In her practice she was adept at using a variety of modalities, including nutrition, herbs, homeopathy, oral vitamins and minerals, intravenous and intramuscular vitamins and minerals, acupuncture, energy therapies, light therapy and prescription drugs.

Dr. Dean met Dr. William Crook in 1986 when they appeared together on a popular Toronto television show. At that point she had already been treating patients with yeast overgrowth for several years, and was invited on the show because she was a well-known local yeast expert. During the 90-minute show, almost 80,000 people called in to find out more about yeast overgrowth!

Now life has come full circle, and Dr. Dean is the medical health advisor for YeastConnection.com. She is honored to help carry on Dr. Crook's monumental work, along with his daughter, Elizabeth Crook, and to share her considerable knowledge with people who want to take responsibility for their own health.

Dr. Dean has lived in New York since 1992, researching medical modalities that can treat layers of infection in the body, including yeast. This research has resulted in a program she developed called the "Body Rejuvenation Cleanse," which will soon be available to the public.

Dr. Dean has published ten books, with four more on the way: *When You Can't Reach the Doctor; Menopause Naturally; Homeopathic*

Remedies for Children's Common Ailments; Natural Prescriptions for Common Ailments; The Miracle of Magnesium; Everything Alzheimer's; Death by Modern Medicine; Hormone Balance; IBS for DUMMIES and *The Complete Natural Medicine Guide to Women's Health.* She has also contributed chapters, sections and quotes to countless other books and publications.

Dr. Dean is a medical advisor to *Natural Health* magazine and a regular guest on TV and radio. She has appeared numerous times on ABC television's *The View* as well as on Fox News, CBS and NBC television.

Dr. Dean also co-hosts two radio shows. On Wednesdays from 9 to 10 A.M. she is on *Lookin' Out For You,* broadcast live on WNTI radio in Hackettstown, NJ as well as over the Internet at www.wnti.org. On Sundays from 7 to 9 P.M. you can listen to *Building Organic Bodies* on Home Grown Radio in New Jersey, where she discusses media, medicine and meditation with other hosts. You can also hear that show at www.buildingorganicbodies.com.

You can visit www.carolyndean.com to learn more about Dr. Dean and her many activities.

Elizabeth Crook

Elizabeth Crook, daughter of Dr. William G. Crook, has spent her career helping people and organizations "take charge" and create positive futures for themselves. She has worked in corporate and not-for profit environments in the U.S., Canada and Latin America.

Since 1992, she has been principal of Elizabeth Crook & Associates, Advisors and Strategists to Entrepreneurs and Organizations. This successful business consultancy helps organizations, entrepreneurs and individuals channel *their* experience and knowledge in order to make a difference in the world. Clients range from large institutions to fast-growth companies to foundations and boards. In addition to being a skilled facilitator, Elizabeth is a sought-after speaker on strategic growth, managing change, and personal strategic planning. For more than four years, she has served on the faculty of the Social Venture Institute at Hollyhock in British Columbia.

Her lifelong interest in health, and especially women's health issues stems from her earliest role model and mentor, her father, and their endless hours of discussion of health issues.

Currently she works on behalf of women's health issues as a consultant to physicians, nurse practitioners, nutritionists and other health care professionals. She is also actively involved on a volunteer basis with several non-profit organizations in Nashville, including an integrative healthcare center.

Elizabeth is a graduate of Vanderbilt University and holds an M.S. degree from Tennessee State University.

For more information, visit her website: www.elizabethcrook.com.

Index

Index

Accutane (isotretinoin), 156
acidophilus douche, 67
acne medication, 22
ALCAT test, 26
alcoholic beverages, 199
Allan Magaziner Center for Wellness and Anti-Aging Medicine, 46
allergies. See chemical sensitivities; food allergies; food sensitivities; mold sensitivities
American Academy of Allergy, Asthma and Immunology (AAAAI), 33–35, 43
American Academy of Environmental Medicine, 44–45, 134
American Association of Naturopathic Physicians, 44
American Autoimmune Related Diseases Association, 160
American College for Advancement in Medicine, 45
American Fibromyalgia Syndrome Association, 119
American Holistic Medical Association, 45
antibiotics: and vaginitis, 64; and yeast connection, 12, 16–17, 20, 21, 22, 177–178; in our food supply, 251
antifungal medication, 38–39, 41, 191–192, 210–217; for endometriosis, 75; for PMS, 59; for sinusitis, 182–183. See also Diflucan; miconazole; Nizoral; nystatin; Sporanox
antioxidants, 241
aromatherapy, 61

arthritis. See psoriatic arthritis; rheumatoid arthritis
Arthritis Foundation, 160
artificial sweeteners, 199
Ashford, Nicholas A., 135
aspartame, 199
asthma, 174; symptoms of, 174–175; treatment for, 175; and yeast connection, 175–180
Atkins, Robert, 158
atopic dermatitis, 14
autoimmune disorders, 151–161; websites relating to, 160–161; and yeast connection, 156–160, 248
azole drugs. See Diflucan; miconazole; Nizoral; Sporanox

bacterial vaginosis, 65–66
Baker, Sidney M., 18–19
Ballweg, Mary Lou, 72–73, 78, 135
Bennett, John E., 49
berberine, 224, 225
Berkson, D. Lindsey, 136
beverages, 198–199
BioMeridian testing, 173
birth control pills, 13, 21
bladder problems. See interstitial cystitis
Bland, Jeffrey S., 225
boric acid suppositories, 67
Braun, Bob, 48
breads, 197
Brodsky, James H., 14–15, 50, 142, 223
Brody, Jane, 170–171
bromelain, 242

New Information For The Twenty-First Century!

THE YEAST CONNECTION HANDBOOK

How Yeasts Can Make You Feel
"Sick All Over"
and the Steps *You* Need
to Take to Regain *Your* Health

William G. Crook, M.D.

THE YEAST CONNECTION COOKBOOK

A Guide to Good
Nutrition and Better Health

A Companion to
"The Yeast Connection and the Woman"
and
"The Yeast Connection Handbook"

William G. Crook, M.D.
&
Marjorie Hurt Jones, R.N.

TIRED-SO TIRED!

and the
"yeast connection"

WILLIAM G. CROOK, M.D.

with a foreword by Bernard Rimland, Ph.D., Director, Autism Research Institute

NOT ONLINE?

Call, fax, or write us to learn more about other books and new products that can help you take charge of your health.

PROFESSIONAL BOOKS
1-800-241-8645
FAX 731-660-5029
P.O. BOX 3246
JACKSON, TN 38303
www.yeastconnection.com

To Order:
- *The Yeast Connection Cookbook*
- *The Yeast Connection Handbook*
- *Yeast Connection Success Stories*
- *Tired So Tired!*
- *Help for the Hyperactive Child*

Call **Wellness Pharmacy**
1-800-227-2627